INTELLIGENT INFORMATION RETRIEVAL:
THE CASE OF ASTRONOMY AND RELATED SPACE SCIENCES

ASTROPHYSICS AND SPACE SCIENCE LIBRARY

A SERIES OF BOOKS ON THE RECENT DEVELOPMENTS
OF SPACE SCIENCE AND OF GENERAL GEOPHYSICS AND ASTROPHYSICS
PUBLISHED IN CONNECTION WITH THE JOURNAL
SPACE SCIENCE REVIEWS

VOLUME 182

INTELLIGENT INFORMATION RETRIEVAL: THE CASE OF ASTRONOMY AND RELATED SPACE SCIENCES

Edited by

A. HECK

Observatoire Astronomique, Strasbourg, France

and

F. MURTAGH

ESO ST-ECF, Garching bei München, Germany

KLUWER ACADEMIC PUBLISHERS

DORDRECHT / BOSTON / LONDON

Library of Congress Cataloging-in-Publication Data

Intelligent information retrieval : the case of astronomy and related space sciences / edited by A. Heck and F. Murtagh.
p. cm. -- (Astrophysics and space science library ; v. 182)
Includes index.
ISBN 0-7923-2295-9 (alk. paper)
1. Astronomy--Data bases. 2. Space sciences--Data bases.
3. Information storage and retrieval systems--Astronomy.
4. Information storage and retrieval systems--Space sciences.
5. Expert systems (Computer science) I. Heck, A. (André)
II. Murtagh, Fionn. III. Series.
QB51.3.E43I56 1993
025.06'5005--dc20 93-19301

ISBN 0-7923-2295-9

Published by Kluwer Academic Publishers,
P.O. Box 17, 3300 AA Dordrecht, The Netherlands.

Kluwer Academic Publishers incorporates
the publishing programmes of
D. Reidel, Martinus Nijhoff, Dr W. Junk and MTP Press.

Sold and distributed in the U.S.A. and Canada
by Kluwer Academic Publishers,
101 Philip Drive, Norwell, MA 02061, U.S.A.

In all other countries, sold and distributed
by Kluwer Academic Publishers Group,
P.O. Box 322, 3300 AH Dordrecht, The Netherlands.

Printed on acid-free paper

Printed in the Netherlands

"It is the destiny of astronomy to become the first all-digital science."

L. Small, in *The Economist*, October 17, 1992.

Table of Contents

Preface 1
B. Hauck

1. Introduction 3
A. Heck and F. Murtagh

Enabling Technologies

2. Understanding and Supporting Human Information Seeking 9
N.J. Belkin

3. Advice from the Oracle: Really Intelligent Information Retrieval 21
M.J. Kurtz

4. Search Algorithms for Numeric and Quantitative Data 29
F. Murtagh

5. Information-Sifting Front Ends to Databases 49
H.-M. Adorf

6. What Hypertext can do for Information Retrieval 81
R. Bogaschewsky and U. Hoppe

Wide Area Network Resource Discovery Tools

7. Archie 103
A. Emtage

8. WAIS 113
J. Fullton

9. The Internet Gopher 119
F. Anklesaria and M. McCahill

10. WorldWideWeb (WWW) 127
B. White

State of the Art in Astronomy and Other Sciences

11. Information in Astronomy: Tackling the Heterogeneity Factor 135
M.A. Albrecht and D. Egret

12. Multistep Queries: The Need for a Correlation Environment 153
M.A. Hapgood

13. Intelligent Information Retrieval in High Energy Physics 173
E. van Herwijnen

14. Astronomical Data Centres from an IIR Perspective 193
M. Crézé

15. Epilogue 197
A. Heck and F. Murtagh

Abbreviations and Acronyms 199
Index 207

Preface

Bernard Hauck

President, International Astronomical Union Commission 5, "Documentation and Astronomical Data"

Director, Institut d'Astronomie
Université de Lausanne
CH-1290 Chavannes-des-Bois (Switzerland)
Email: hauck@cgeuge54.bitnet

Astronomers are the oldest data collectors. The first catalogue of stars is due to Hipparchus, in the second century B.C. Since that time, and more precisely since the end of the last century, there has been an important increase in astronomical data. Due to the development of space astronomy during recent decades, we have witnessed a veritable inflation.

Confronted with this flood of data, astronomers have to change their methodology. It is necessary not only to manage large databases, but also to take into account recent developments in information retrieval.

Astronomy and space sciences are now evolving in a context where information is a combination of text, numerical data, graphics and images. All these aspects can of course be combined in a single scientific paper, and are also increasingly retrieved and amalgamated from differently structured databases (such as SIMBAD) and archives (such as those of Hubble Space Telescope) located in various places around the world and linked through specific networks. The *Centre de Données* of Strasbourg plays in this context a primordial role. As one of the first foreign participants in this centre, I am impressed by its evolution during the last twenty years and by the excellent foresight of the initiators.

Databases and archives are used both upstream and downstream of actual scientific research and publication of papers. Astronomers and space scientists are now moving towards electronic publishing, the final aim of which can be considered as being what is called "intelligent" information retrieval.

The present book is definitely timely. Its chapters cover various aspects of IR and well meet the scope of the Commission on *Documentation and Astronomical Data* of the International Astronomical Union.

The editors' intention was to produce an educational book with emphasis on astronomy and space sciences but it will be quite profitably read by other scientists too. For subjects outside the realm of astronomy and space sciences, the editors have ensured first-rate contributions from outstanding specialists.

Surely the best guarantee of the book's quality is the experience of the editors. André Heck has had many astronomical interests, his main involvements now being electronic publishing and intelligent information retrieval. André Heck is the IAU representative to CODATA. Fionn Murtagh is a senior scientist at the European Space Agency. His first interests were statistics and information retrieval, his scientific interest now being concerned with the interface between astronomy and astrophysics, on the one hand, and computer sciences and statistics on the other. Together the editors have been organizers of several conferences and editors or authors of a number of books. Both can be considered as leaders in the subject and we can hope that the present volume will play an important role in a field that is in full evolution.

Chapter 1

Introduction

André Heck

Observatoire de Strasbourg
11, rue de l'Université
F-67000 Strasbourg (France)
Email: u01105@frccsc21.bitnet

and

Fionn Murtagh[1]

Space Telescope – European Coordinating Facility
European Southern Observatory, Karl-Schwarzschild-Str. 2
DW-8046 Garching/Munich (Germany)
Email: fmurtagh@eso.org

To avoid any ambiguity, it is important to make clear from the beginning what the concept of *information* covers in astronomy and space sciences. Generally speaking, it embraces textual, numerical and pictorial data. The latter could be graphics, pictures or satellite images. Numerical data could be tabulated. Text could include a non-negligible amount of mathematical formulae.

Retrieving information concerns not only classical bibliographical references, but also abstracts and scientific contents of papers, as well as tabular data, subsets of catalogues, and observational material from ground-based and space-borne instrumentation.

Intelligent information retrieval (IIR), or better *advanced* information retrieval (AIR), implies added flexibility or additional degrees of freedom relative to information retrieval. For instance, one may take advantage of the fact that the material to be handled is in machine-readable form.

[1]Affiliated to the Astrophysics Division, Space Science Department, European Space Agency.

A. Heck and F. Murtagh (eds.), Intelligent Information Retrieval: The Case of Astronomy and Related Space Sciences, 3–8.

A loose definition of IIR could be the following: one does something more elaborate than just looking for the presence of a string of characters in a base and retrieving the data attached to it.

The concept takes its full meaning with sophisticated algorithms that can be put in parallel with two aspects of human thought: *logic* and *association*. The latter applies to ideas, concepts, terminologies, and so on, even if the relationships involved are sometimes implicit and/or subtle. The same philosophy can be found also in a hypertext approach or in techniques such as artificial intelligence (AI) or expert systems.

The potential intelligence of a retrieval is conditioned firstly on the type of information that is stored; secondly, by the way the database is structured (database engine, user interface, and so on); and finally by the tools used to search it and by the range of algorithms constructed around the database. A significant current trend in databases is to shift towards the more general concept of *knowledge* bases.

Searching for textual information in these bases will call for thesauri, the construction of which is far from obvious and immediate. They must take into account hierarchical or relational links between the terms and expressions listed (synonymities, analogies, broader or narrower concepts, and so on). They will have to be kept up-to-date, as science progresses and evolves. Ad hoc expertise will have to be secured. There is a risk that thesauri will always lag behind.

Adequate and homogeneous indexing of documents will have to be performed in parallel. We cannot but agree with Locke (April 1991 issue of *Byte*) when he writes: "Indexing documents properly and consistently for later retrieval is not a low-order clerical task, but a complex exercise requiring knowledge engineering skills". This aspect, which is currently grossly underestimated, will have to be carried out in collaboration with well-trained cognitive scientists, once the languages used by the various parties are tuned to each other.

Hypertext systems have also been investigated as means for storage, retrieval and dissemination of information. Hypertext can be characterized as a document allowing, contrary to classical texts, non-sequential and multiple *reading* paths, selected by the consultant, which implies an interaction between this person and the infrastructure.

The navigation along these paths is regulated by the principle of idea associations (analogies), sometimes with some statistical background, but can also be piloted by situations. There are more logical capabilities in it than in a sequential reading (classical text is not always sequential or linear: footnotes and references break this flow) or in a hierarchical database.

Hypermedia systems add media facilities, such as displays, animated graphics, digitized sound, voice, video, and so on. Hypermedia presentations combining all these facilities are really impressive.

Already about a quarter of a century ago, the basic problems in documentation were the same as those heard today: too much publishing, too many papers, means of retrieval requiring computerization, selective dissemination of information (SDI), need for developing advanced algorithms to get quickly the relevant material out of the information jungle, and so on.

Since the quantity of papers published roughly doubles every 15 years (currently some 10^9 papers/year), the situation has worsened in terms of bulk of publications and original volume of paper. The difference is that, a quarter of a century ago, the medium for encoding information was almost exclusively paper via punched cards or tapes, with a monotask computer occupying a whole basement and itself generating punched cards for intermediary steps. The power of such a computer is now superseded by a basic A4-size laptop notebook, not to speak of the stacks of boxes of punched cards replaced by a handful of diskettes or a few tracks on a PC hard disk.

The communication of scientific results, and of astronomical ones in particular, which was essentially carried out up to now on paper (sometimes as an extension of an oral presentation), is currently probing other means such as electronic ones. As a consequence, will a new deontology take shape in the corporation?

The motivations and modalities of communicating are nowadays basically oriented towards the need for recognition and this in turn, essentially through traditional publishing (implying subsequent IR), for getting positions (grants or salaries), acceptance of proposals (leading to data collection) and funding of projects (allowing materialization of ideas). This situation is likely to remain the same as long as the *publish or perish* rule conditions publication policy.

Most of the underlying rules were designed at a time when the pen (and then the typewriter) and noisy telephone lines over manned switchboards were the basic tools for communicating and preparing documents to be printed. At the same time, the index card, when it was there at all, was the only access key to a possible archive or to the elements of a library. The only medium was the paper and the constitution of libraries relied heavily on classical mail, depending itself on the means of transportation of that epoch.

Things have substantially changed since then. Paper, which is still present and is likely to remain so, is far from being the only infrastructure for information: microfiches and microfilms were fashionable some time ago and have not yet disappeared. Magnetic media of all kinds (tapes, cassettes, floppy and hard disks, and so on) are now omnipresent. Video-optical devices are in full expansion and already used by astronomical journals.

The computing revolution is far from being completed. It gained new dimensions in the last decade with the introduction of personal computers and workstations, more and more powerful, as well as by the availability of ever more sophisticated text processing systems and the spread of data networks, the popularity of which reflects their intrinsic usefulness in our present society.

Laser printers produce high-resolution pages of such top quality that it is sometimes difficult to distinguish, when receiving a document, whether this is a preliminary draft, an intermediate working step or a final product coming from a publisher with the blessing of all parties involved. This is not without raising a few problems as the so-called *grey literature* is more and more difficult to identify.

Data, information and/or knowledge bases are intimately involved with our activities: observing (remote or traditional), data reduction, general research, publication, management of instrument time, and so on. The knowledge retrieved from them will be a combination of text, quantitative and qualitative data, n-dimensional raw and/or reduced observing material, as well as tabular and pictorial information in general.

As indicated earlier, the means of retrieval have substantially evolved since the time the index card was the only access to library shelves and to the *Astronomischer Jahresbericht* (now *Astronomy & Astrophysics Abstracts*) volumes, the privileged gate to specific papers, indexed essentially through keywords and author names.

At the beginning of the 1970s, a major step forward was taken when the Strasbourg Data Centre (CDS) initiated the enormous task of compiling both an object cross-identification table called the Catalogue of Stellar Identifications (CSI) and the Bibliographical Star Index (BSI) where papers were indexed through all objects mentioned in them (in the whole text, not only in the title or in the abstract).

It was then feasible, from one object identification, to access all others and the corresponding data from different integrated catalogues, as well as the bibliography mentioning this object under its various denominations. Later on, CSI, BSI, and the catalogues were integrated into an on-line database called SIMBAD used now worldwide (and *stellar* was replaced by *Strasbourg* in the acronyms because of the opening of the database to non-stellar objects).

Databases and archives (essentially of space missions) have been multiplying in the last decades. In its lifetime, Hubble Space Telescope will produce an estimated 15 Terabytes of data, and the associated archival research will open unprecedented challenges. There are projects (such as ESIS and ADS) aiming at linking the major databases and archives through networks and at offering standardized (as far as possible) user-friendly interfaces.

New tools such as computerized faxes, scanners, OCR packages and cheap postscript printers have become extremely useful in handling documents in a machine-readable way. Distributed, wide-area information servers have become very popular, as well as exploders and news groups. FITS, a standard designed by astronomers, has now been adopted in other communities for file transfer and handling.

Institutions are setting up their own servers for all kinds of purposes: catalogues, preprints, proposals, documentation, macros for electronic publishing (EP), and so on. EP itself will be a powerful tool, interacting with knowledge bases upstream and downstream.

New means of communication are eliminating geographical separation and have revolutionized the way scientists transfer information to each other. Now, through networks and remote logon, scientists have direct access to remote databases and archives, and can retrieve images or data they can process locally, as well as tables of data they can directly insert in a paper to be published. Conversely, once it is accepted by a journal after appropriate screening and quality control, a compuscript can be channelled into a database from which it could be retrieved even before being published on paper, if ever. Its bibliographic information (at least its general references, the keywords, the abstract, its quotations) can be accessible immediately through the existing bibliographical databases and citation indices.

Subject to the design of specific procedures and techniques for plots, graphs, figures and pictures, even the whole paper can be retrieved. Catalogues, extensive lists of data or tabular data, can also be channelled through networks into ad hoc bases such as SIMBAD.

It should be noted that the journal *Acta Astronomica* maintains an anonymous ftp account from where detailed unpublished observational data can be retrieved. Such a system is definitely less cumbersome than writing to a data centre where people would have to photocopy the material on paper or duplicate a magnetic tape (or diskette) and send it through the post. *Astronomy & Astrophysics Supplement Series* has entered a similar joint venture with CDS. In fact, more and more references are now given together with their availability from anonymous ftp accounts, and are sometimes not accessible otherwise.

It will be of fundamental importance to involve librarians in the future developments and to take their standpoints and requirements into account. They, too, will have to participate in the modernization of knowledge search in their realms. Surveys have already been carried out in some institutions. New techniques such as bar coding have been implemented. It is clear that we have entered a new age where librarians have a new attitude towards IR and where scientists have a new attitude towards their librarians.

Once again, it seems obvious that the combination of desktop and electronic publishing with networking and the new structuring of knowledge bases will profoundly reshape not only our ways of publishing, but also our procedures for communicating and retrieving information.

Generally evolution is accompanied by quite a few surprises and it would be dangerous – and pretentious – to play here the game of guessing what they will be. What is sure is that technological progress will play a key rôle in future orientations. But will there be forever a need for publishing which, as such, will become increasingly integrated with data reduction and analysis? There might be a time when the keyboard will not be necessary anymore and when the journals as we know them today, or even publishing as such, will disappear, to be replaced by some electronic form of dialogue with the machine and knowledge bases.

Direct computer pen writing (implying handwriting recognition) is already taking over from mouse clicking and pointing. Gesture and voice recognition will be common in the near future. The idea of direct connection between brain and computer is not new. It is used, among others, by cyberpunks. Their cyberspace, or knowledge space related to what is also called virtual reality, is where the brain navigates with a virtual body and has access to all kinds of collections and sets of data. Cyberpunks have sometimes so much puzzled government officials with their innovative – often imaginary – use of computer techniques and networking that some publishing companies have been raided by secret services.

Part of these last comments is still fiction, but we might expect with Jules Verne that *Tout ce qu'un homme est capable d'imaginer, d'autres hommes seront un jour capables de le réaliser* (Everything that man is able to imagine will be achieved later by others). As things are changing rapidly, there must be a readiness from the scientific community to go along with them and get involved in their implementation, essentially at a time when resources for basic or fundamental research seem to be rapidly drying up.

The chapters of this book deal in a tutorial approach with most of the various points listed above.

It is a pleasant duty to acknowledge here the cooperation of contributors to this book for the timely delivery of their respective chapters. It was also a pleasure to collaborate with the publisher on the production of this volume.

Chapter 2

Understanding and Supporting Human Information Seeking

Nicholas J. Belkin

School of Communication, Information and Library Studies
Rutgers University
New Brunswick, NJ 08903 (USA)
Email: belkin@zodiac.rutgers.edu

Abstract

What would constitute "intelligent" information retrieval is a question which has had a variety of answers over the twenty or so years that the concept has been extant. This paper reviews some of the approaches to intelligent information retrieval that have been suggested. Most of these suggestions have shared the assumption that the "intelligence" resides solely in the built system. More recent work in information retrieval has begun to concentrate on the interaction between the user and the other components of the system, and this work leads to an alternative assumption: that the user is an integral part of the information retrieval system, and that intelligence in the system resides in appropriate distribution of tasks between the user and the built system. We review some of this newer work, and propose an approach to the design of information retrieval systems starting from this new assumption. The approach is based upon three principles: that information retrieval is a process in support of the user's more general domain and task environment; that people engage in a variety of information seeking strategies, all of which should be supported by the information retrieval system; and, that information retrieval is an inherently interactive process among user, knowledge resources, and intermediary mechanisms. These principles lead to prescriptions for domain, work and task analysis, for study of information seeking strategies and their relationships to underlying tasks, and for system design to support cooperative interaction among the participants.

A. Heck and F. Murtagh (eds.), Intelligent Information Retrieval: The Case of Astronomy and Related Space Sciences, 9–20.

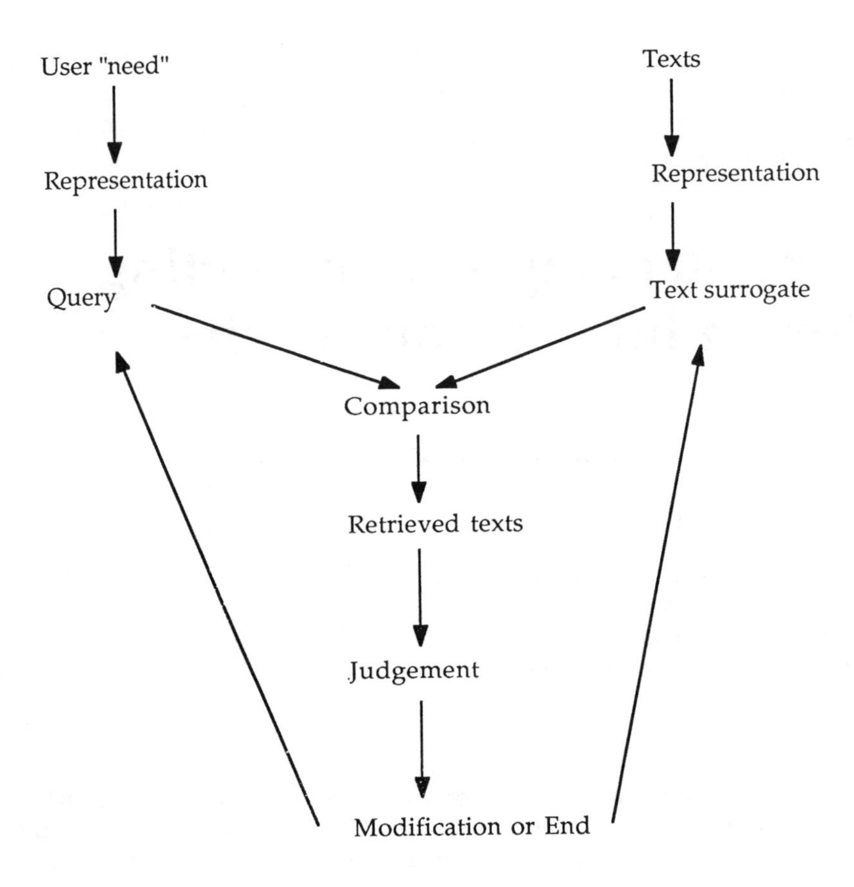

Figure 2.1: A model of information retrieval.

2.1 Introduction

Before beginning to speak of "intelligent" information retrieval, it is relevant to say a few words about what ordinary, "unintelligent" information retrieval is thought to constitute. The most typical view of information retrieval (IR) is that it is a process whose role, and goal, is the retrieval of texts which are relevant to a user's query, put to an IR system. In this sense, IR has been concerned primarily with textual information, although a text, in principle, is construed as any information-carrying object (so that multi-media texts are natural objects in IR systems). From this point of view, IR can be taken to be represented by the model displayed in Figure 2.1.

IR, as in Figure 2.1, consists of several processes operating upon, and producing, a number of entities. It begins with a person with a so-called "information need", who instigates the system. This need, or problem, then is subjected to a process of representation, leading to a formalized query. On the other side of the diagram, the focus is on the texts to which the person (commonly referred to as the "user") desires access. These texts also undergo a representation process, leading to a set of text surrogates. This process, in current systems, is usually some form of indexing, and its result is some indexing statement, such as a set of keywords. These surrogates are typically organized according to some scheme, such as a classification, or an inverted index. The query is then compared to the set of text surrogates, in order to retrieve texts which are likely to be relevant to the information need. These retrieved texts are judged, in principle by the user, according to their relevance to the information need. The user is either satisfied, and leaves the system, or some modification is made of the query (this process is known as relevance feedback), or, more rarely, of the text surrogates (known as document learning), and the system iterates from the comparison process. Although more complex models of IR have been proposed and used as the basis for research and system design (Belkin and Croft, 1992; Ingwersen, 1992), this model constitutes a "standard" view of IR. IR from this standard point of view has thus been especially concerned with processes of representation of text and information need, with comparison between query and text surrogates, and with relevance judgements and the modification of query representations. Research within this paradigm of IR has been going on for at least 30 years, and has lead to several interesting, and generally consistent results.

First, comparison processes based on "best match" principles, which rank the texts in the database according to some estimate of relevance of text to query, provide better performance than those based on "exact match" principles, which partition the database into two sets, relevant and non-relevant. The major reason for this result is presumed to be that the representations of both information need and texts are inherently uncertain, for a variety of reasons (van Rijsbergen, 1986; Furnas et al., 1983), and that exact match comparison is therefore inherently unduly restrictive. However, using the structure associated with Boolean query formulations (typically associated with exact match retrieval) in a best match comparison process has been shown to be effective (Salton et al., 1983; Turtle and Croft, 1991). Second, text representations based upon the language of the texts themselves provide results equal or superior to text representations based upon controlled vocabularies, such as thesauri. Since both the construction and application of controlled indexing vocabularies are highly labor-intensive, and thus expensive, this result has led to the conclusion that automatic methods based on the texts themselves are better; that is, more cost-effective. Third, the overall performance of almost any system which is sensibly designed (this means at least with best match comparison), no matter what the specifics of the representation and comparison processes, will be about the same as that of any other system, given the same caveat. Of course, there will be differences in the performance of different systems, but such differences may not be consistent from database to database, and in any event are usually fairly small. An extremely interesting aspect of this result is that although the performance is roughly the same, the actual documents which are retrieved (both relevant and non-relevant) by different systems, are different (Saracevic and Kantor, 1988; Foltz and Dumais, 1992; Harman, 1993). A final general result of IR research is that the most effective way to improve IR system performance is by incorporating relevance feedback; that is, by building interaction between user and the rest of the system into the IR system design explicitly (Salton and Buckley, 1990).

It is important, when discussing the results of IR research, to note the criteria and conditions under which these results are obtained. The evaluation criterion for all of the results cited above has been the relevance of texts to a specified query; the performance measures operationalizing this criterion have been recall (the proportion of texts in the database which are relevant to the query which have been retrieved) and precision (the proportion of retrieved texts which are relevant). These correspond to the intuitive assumption that optimal performance of an IR system is to retrieve all, and only, the relevant texts in the database. Clearly, the recall measure requires knowledge of which documents in the database are relevant to any given query. This has led to the most important condition of classical IR research; that is, that it is carried out in the so-called "test collection" environment. A test collection consists of a database (usually of text surrogates, such as abstracts), a set of queries posed to that database, and an exhaustive set of relevance judgements for each query. Then, an experiment is conducted by comparing the performance of different techniques (representation or comparison) averaged over the set of queries, in terms of recall and precision. Such test collections must, of necessity, be small, since the problem of judging all texts in the database against all queries, is so highly labor-intensive.

This experimental and evaluative methodology is certainly constraining, and may even lead to misleading results. Some obvious problems that arise include the following. First, the extent to which the measures are realistic is at least questionable. One can imagine, for instance, many classes of information problem for which having only one relevant document would be better than having all relevant documents, and, conversely, situations in which the user will be relatively insensitive to how many retrieved texts it is necessary to look through before finding a (or some) relevant ones. Second, there is an obvious question about whether the results in such small collections (ranging from 600 to, at most, 10,000 texts) and with so few queries (almost always less than 100) are scalable to the situations of real databases (of hundreds of thousands, or millions of texts), and generalizable to the universe of information problems. Third, the specific experimental context almost completely excludes the possibility of investigating the effect of interaction between user and the rest of the system. All of these issues have been brought up throughout the history of experiment in IR, but it is only fairly recently that there have been substantial efforts (experimental and theoretical) to address them (Ingwersen, 1992; Harman, 1993; Su, 1992). So, although there is a substantial body of results from traditional IR research, there seem to exist valid reasons for questioning the applicability of at least some of them to the realistic, operational environment of the interaction of a great variety of users with very large databases, often consisting of full texts. These issues, as well as other theoretical concerns, have lead to what we can construe as attempts at the design of intelligent IR systems.

2.2 Intelligent Information Retrieval

So-called "intelligent" IR has typically been taken to mean the development of computer-based IR systems which will perform the processes of the standard IR model automatically and with the use of some sort of "knowledge", or which will perform some of the functions which take place in IR systems which are normally done by human beings (Croft, 1987). Examples of such functions include those of human intermediaries, such as interacting with the user to specify the information request, selecting appropriate data bases, developing and modifying the search strategy, and explaining the system to the user. Usually, it has been assumed that such a system will be designed and built in a knowledge-based or expert system paradigm. Automatic indexing, perhaps based on natural language analysis, is an example of intelligent performance of one of the IR processes. Other views of what has been termed intelligent IR have focused on the system's being able to do things automatically that humans have been unable to do, or to do well, because of constraints such as time or size, such as linguistic processing of large amounts of text, or finding correlations among large numbers of documents.

This approach constitutes what one might now call the "traditional" view of intelligent IR. Whatever the specifics of the particular proposal, all of them within this paradigm have depended upon three somewhat problematic assumptions. The first is that intelligence in the system resides solely in the built IR system. Related to this assumption is the assumption that the user is not a part of the IR system, but is rather external to it. And the third is that one can treat the IR task in general, separately from the user's overall goals and task environment. These assumptions have lead to systems and theories which have difficulties accounting for interaction between the user and the other components of the IR system; whose evaluation criteria often don't respond well to users' goals, and which are typically rather far removed from the users' normal activities.

Despite the limitations imposed by these unrealistic assumptions, there has been a good deal of significant work done in the general area that goes beyond the standard IR paradigm. For instance, since the late 1970s, one can trace a line of research in which the underlying concept of IR is inherently interactive (Oddy, 1977), the user is considered a central component of the IR system (Belkin et al., 1983), models of the user and of the user's situation are constructed interactively and used dynamically (Brajnik et al., 1987; Daniels, 1986), and the user's own knowledge is elicited and used by the other components of the system (Croft and Thompson, 1987). Examples of other research efforts which can legitimately be included under the rubric of intelligent IR include the use of probabilistic inference nets as an underlying technique for incorporating multiple sources of evidence in retrieval models (Turtle and Croft, 1991), basing interaction between system and user on a model of belief revision (Cawsey et al., 1992), and designing IR system interfaces to support highly flexible interaction (Belkin et al., 1993a). There is also an increasing amount of activity in the application of concepts from connectionist and neural network research to the problems of IR, for instance in the question of document learning (Belew, 1989). Furthermore, initiatives such as the Text Retrieval Conference series (Harman, 1993), have been established explicitly to respond to some of the problems with test collections and evaluation of system performance that were discussed above. All of this work is leading to rather new and interesting IR systems which do not depend completely upon the restrictive assumptions of the traditional IR paradigm. Nevertheless, none of them is entirely free of all of these assumptions, and most still do not address the question of the match between IR system goals, and user's overall goals.

In order to respond to such problems, and, indeed, to deal more realistically with the IR situation, we suggest that an alternative concept of intelligent IR be developed, based on a set of alternative assumptions about the nature of the IR situation.

2.3 An Alternative Concept of Intelligent Information Retrieval

The assumptions upon which an alternative idea of intelligent IR could be based are: that the user is an active participant in the IR system, rather than only a passive user of, or respondent to it; that intelligence resides in both the user and the built system (especially in appropriate distribution of tasks between these two parties); and, that IR is a tool in support of the user's more general task environment.

These alternative assumptions about the nature of IR in general lead to alternative views of what might constitute intelligent IR. Roughly, from this point of view, we might say that an intelligent IR system is one in which: user and system cooperate in the IR task; user and system cooperate in achieving the domain task or goal, through IR interaction; and, the IR tasks themselves are distributed between the user and the rest of the system appropriately. This is a rather tall order. Here, we only suggest what we believe needs to be done in order to build such systems, describe one general approach to this problem, and give a few examples of how these design tasks might be accomplished.

The first step in designing an intelligent IR system in this model, is to perform a domain, or work, analysis. By this is meant obtaining some representation of the field of the potential users of the IR system, in particular their goals, and how they go about achieving them, in their daily setting. This is required in order to have a context in which to embed the IR system, and to relate its tasks and output to the users' overall environment. A good example of this type of analysis, at one level, is the domain model underlying the ESIS Query Environment (Giommi et al., 1992).

The second step, related to the first, is to perform an analysis of the tasks undertaken by the people in the specified domain. Such a task, or work analysis, will relate the goals of the people in the domain to activities taken by them in order to achieve those goals. A crucial aspect of this type of task analysis is the identification of problems that people face in accomplishing their tasks, as well as what they do and use to overcome the problems. The function of a task analysis (more explicitly, a Cognitive Task Analysis: Woods and Roth, 1988) is to identify specifically what demands the IR system must respond to, in terms of the users' requirements and constraints. The interviewing methods that were used in developing the ESIS Query Environment illustrate one means to accomplish task analysis of this sort (Giommi et al., 1992). A task analysis at this level, is concerned with the users' normal working environment. The next step in the IR system design process is to relate the results of this analysis to the activities that need to be undertaken by, and within the IR system.

In order to relate overall tasks to the IR system, it is necessary first to identify and characterize the knowledge resources required for accomplishing the users' tasks. This specifies, for instance, the content and structure of the databases of the IR system. These can be either existing databases, or this analysis could lead to the construction of new databases, or of new combinations of existing ones. The support of the user in interaction with the databases and other components of the system is based on the next stage of analysis, the cognitive task analysis for the domain of IR in general, and for information seeking activities in the specific task environment. Belkin et al. (1993a) describe such an analysis for the user in the IR system.

A slightly more general task in the design of intelligent IR systems is the characterization of information seeking behaviors. By this is meant identifying what people in the domain do when they search for information, from whatever source, why they engage in each particular behavior, when, and with what success. This kind of characterization will lead to a specification of the range of information seeking activities which the IR system will need to support, and how they relate to one another. This kind of characterization provides a basic framework for interaction in the IR system.

Having accomplished these tasks, it is then necessary to establish relationships among the tasks and goals, the information seeking behaviors, and the knowledge resources. These relationships are of the type indicated in Figure 2.2. The domain and task goals are related to specific information seeking intentions, which can be accomplished by a variety of information seeking behaviors or strategies, operating upon a variety of knowledge resources. Such relationships are both empirically and logically founded, and lead to a means for specifying appropriate IR system support mechanisms in specific circumstances.

Next comes the task of specifying the appropriate roles for the user, and for the rest of the system. This requires the previous analyses as input, but also a concept of what the users are good at doing (this often means asking the question of what knowledge they have, that the rest of the system cannot have) and what the rest of the system is good at, and of what responsibility each party has. For instance, it is the role of the user to make evaluative judgements of the relevance of texts; it is the role of the other parts of the system to take account of such judgements in order to do other comparisons and suggest other possible texts to the user. Also, it may be the case that the user's knowledge of the domain is more detailed than that of the system. In this circumstance, the user can be expected to offer this knowledge in support of the other representation and comparison processes.

From the results of the previous steps, one can move to a specification of the functional design of the system, which will support the information seeking behaviors which have been identified, within an environment which will couple them to the appropriate knowledge resources (see Belkin et al., 1993a, 1993b, for partial examples of such a specification).

To implement the functional specification, one needs to derive an interface design which will support cooperative, two-party interaction. This is required because of the assumed interaction between user and the rest of the system, and because of the specific

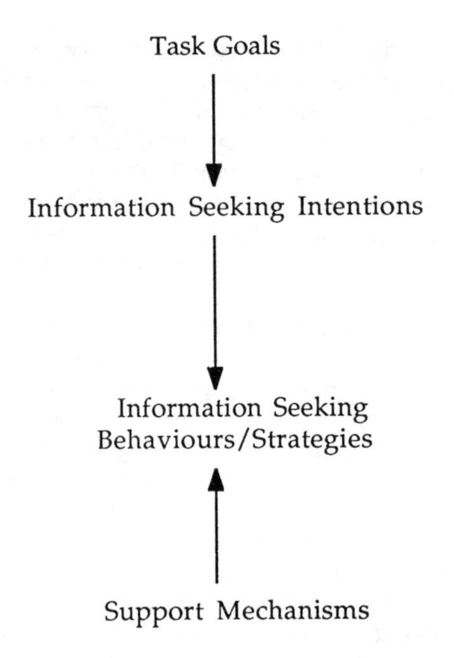

Figure 2.2: Relationships in the IR system design process.

allocation of roles. A reasonable model for such an interaction is the functional characterization of human dialogues in similar settings. Belkin and Vickery (1985) suggest some ways in which such a model could be implemented in an IR environment, and a prototype system based on a formal dialogue structure is described in Belkin et al. (1993b).

Finally, the IR system should be coupled closely to the domain and task environment. In the situation of, say, a working astronomer, this suggests that there be a single environment in the user's workstation, which allows direct access to the IR system from within any application, with relevant results being directly piped to that application, as necessary. A model for this type of coupling is the typical academic office, with shelves of off-prints immediately at hand, to be retrieved and used without leaving the task (say, writing a paper) which is currently being engaged in.

2.4 Accomplishing the Tasks for Intelligent IR System Design

In this section, we wish only to suggest some methodologies by which the tasks identified in the previous section might be accomplished, and to discuss some preliminary results on the characterization of information seeking strategies, and the use of such characterization in IR system design. All of the methods which are discussed below are assumed to take place within a participatory design framework, in which user, designer and builders are all members of the design team, each contributing appropriately to the final design.

In general, task and domain analysis are most appropriately carried out using ethnographic methods. In particular, one would suggest here beginning with participant or non-participant observation of the subject group, moving to some form of elicitation of descriptions of activities, such as structured interviews or focus groups, and then further observation within more controlled settings, such as performance of identified tasks. These same types of methods lead directly to criteria for knowledge resource specification and characterization. This stage of design is necessarily specific to the particular context.

The specification of information seeking behaviors or strategies is also appropriately carried out through ethnographic and other observational methods, but needs also to be corroborated through more controlled investigations, such as predictive experiments in prototype environments. The analysis at this stage could be more general than that of the previous stage, since one might postulate a fairly limited range of such strategies, which are applicable across a wide variety of task domains. The identification of relationships among the previous components is something that can be carried out as a logical exercise, with corroboration being done by comparison against previously collected data, and within experimental frameworks. These relationships are clearly specific to particular task domains, and will need to be established anew for each domain to be supported. The final three steps of the IR system design process are probably best carried out through the process of iterative evaluation-design, in which prototypes are designed, built, tested and re-designed. Egan et al. (1989) is a good example of such a methodology for information system design.

Belkin et al. (1993a) have suggested a means of characterizing information seeking strategies that is relevant to the issue of IR system design. On the basis of some previous studies of people engaged in information seeking, they propose that information seeking strategies can be characterized along several behavioral dimensions, which can be interpreted as defining a space of information seeking strategies. They note that people engage in behaviors that place them in different regions of this space, and that they move from one region to another (that is, they change strategies) according to a variety of locally influencing factors. The dimensions that they tentatively suggest are:

- Method of interaction (scanning vs. searching)
- Goals of interaction (learning vs. selecting)
- Mode of retrieval (recognition vs. specification)
- Resource interacted with (information vs. meta-information)

Thus, browsing in the current journal shelves to see if there is anything interesting to read, could be characterized as scanning, to select, by recognition, in an information resource. These dimensions have been used to design an interface to support free user movement in such a space, when interacting with a large bibliographic information system (Belkin et al., 1993a), and will shortly be used to support interaction in a remote sensing context. It has also been suggested that the regions of such a space might have prototypical interaction sequences associated with them, as dialogues, and a system for supporting such dialogues, and using them to index and use specific previous cases to guide new ones, has been built as a prototype (Belkin et al., 1993b).

In addition to these dimensions and their poles, one might add others, such as a text vs. data dimension, or a discover vs. manipulation dimension. Such dimensions might be general to all of information seeking behaviors, or they might be domain dependent. In any event, investigation of these kinds of behaviors in more domains is necessary, in order to see if there are commonalities across domains. If there are, then we might be able to consider general design principles for the functional specification of an interface to an IR system which allows seamless movement among all of the relevant strategies.

Of course, embedded in this general design methodology framework, we still require techniques for the accomplishment of the basic processes of IR: representation, comparison and modification, and a general structure for implementing these processes in an interactive environment. It is at this level that we can refer to the research that has been done, and continues to be done, in the more usual concept of intelligent IR. In particular, retrieval models based on IR as plausible inference (Turtle and Croft, 1991), concepts for incorporating user models into IR interaction (Brajnik et al., 1987), the use of natural language processing techniques for text and query representation (Liddy and Myaeng, 1992), and models of IR interaction (Ingwersen, 1992), seem especially promising.

Clearly, this methodology for IR system design is complex and time consuming, and requires some results of basic research in human information seeking before it can be completely carried out. But there is already work being done on various aspects of this problem, especially in particular task domains, such as that of astronomy and astrophysics, as the papers in this volume demonstrate. Furthermore, there is now some substantial research, and experience of at least prototype systems based on these principles, in IR. And the potential advantages of such systems, especially in terms of their support of people in attaining their working goals, over more traditional IR systems, seem so great as to justify this level of effort.

2.5 Conclusions

Having said all this, what can we now say about what constitutes an "intelligent" IR system? Such a system would certainly have the following characteristics. It would be a part of the user's normal working environment, with no necessity for the person to leave a task currently engaged in in order to do IR. It would support movement among information seeking and manipulating strategies, and support (or encompass) movement among knowledge resources. It would, furthermore, display results to match the user's goals and intentions, and allow easy incorporation of the results into the task to which they are relevant. In short, it would cooperate with the user, in helping the user to achieve the goal at hand, through providing access to information resources which the IR system judges to be potentially relevant to that goal. But it would leave the final judgement, and the accomplishment of the goal, to the user, where the real intelligence resides. From this point of view, we can say that what we are really attempting to design are not "intelligent" IR systems, but good ones.

References

1. Belew, R.K., "Adaptive information retrieval", *Proceedings of the 12th International Conference on Research and Development in Information Retrieval*, ACM, New York, 11–20, 1989.

2. Belkin, N.J., Cool, C., Stein, A. and Thiel, U., Preprints of the AAAI Spring Symposium on Case-Based Reasoning and Information Retrieval, March 1993b.

3. Belkin, N.J. and Croft, W.B., "Information filtering and information retrieval: two sides of the same coin?", *Communications of the ACM*, **35**, 29–38, 1992.

4. Belkin, N.J., Marchetti, P.G. and Cool, C., "BRAQUE: Design of an interface to support user interaction in information retrieval", *Information Processing and Management*, **29**, 1993a (in press).

5. Belkin, N.J., Seeger, T. and Wersig, G., "Distributed expert problem treatment as a model for information system analysis and design", *Journal of Information Science*, **5**, 153–167, 1983.

6. Belkin, N.J. and Vickery, A., *Interaction in Information Systems*, The British Library, London, 1985.

7. Brajnik, G., Guida, G. and Tasso, C., "User modeling in intelligent information retrieval", *Information Processing and Management*, **23**, 305–320, 1987.

8. Cawsey, A., Galliers, J., Reece, S. and Sparck Jones, K., "Automating the librarian: belief revision as a base for system action and communication with the user", *The Computer Journal*, **35**, 221–232, 1992.

9. Croft, W.B., "Approaches to intelligent information retrieval", *Information Processing and Management*, **23**, 249–254, 1987.

10. Croft, W.B. and Thompson, R.H., "I^3R: a new approach to the design of document retrieval systems", *Journal of the American Society for Information Science*, **38**, 389–404. 1987.

11. Daniels, P.J., "Cognitive modeling in information retrieval – an evaluative review", *Journal of Documentation*, **42**, 272–304, 1986.

12. Egan, D.E. et al., "Formative design-evaluation of Super-Book", *ACM Transactions on Information Systems*, **7**, 30–57, 1989.

13. Foltz, P.W. and Dumais, S.T., "Personalized information delivery: an analysis of information filtering methods", *Communications of the ACM*, **35**, 51–60, 1992.

14. Furnas, G.W., Landauer, T.K., Gomez, L.M. and Dumais, S.T., "Statistical semantics: analysis of the potential performance of keyword information systems", *Bell Systems Technical Journal*, **62**, 1753–1806, 1983.

15. Giommi, P. et al., "The European Space Information System", in A. Heck and F. Murtagh, eds., *Astronomy from Large Databases II*, European Southern Observatory, Garching, 289, 1992.

16. Harman, D., ed., *The First Text Retrieval Conference*, National Institute of Standards and Technology Special Publication 500-207, Gaithersburg, MD, 1993.

17. Ingwersen, P., *Information Retrieval Interaction*, Taylor Graham, London, 1992.

18. Liddy, E.D. and Myaeng, S.H., "DR-LINK: Document retrieval using linguistic knowledge. Project description.", *SIGIR Forum*, **26**, 39–43, 1992.

19. Oddy, R.N., "Information retrieval through man-machine dialogue", *Journal of Documentation*, **33**, 1–14, 1977.

20. Salton, G. and Buckley, C., "Improving retrieval performance by relevance feedback", *Journal of the American Society for Information Science*, **41**, 288–297, 1990.

21. Salton, G., Fox, E. and Wu, H., "Extended Boolean information retrieval", *Communications of the ACM*, **26**, 1022–1036, 1983.

22. Saracevic, T. and Kantor, P., "A study of information seeking and retrieving. III. Searchers, searches and overlap", *Journal of the American Society for Information Science*, **39**, 197–216, 1988.

23. Su, L., "Evaluation measures for interactive information retrieval", *Information Processing and Management*, **28**, 503–516, 1992.

24. Turtle, H. and Croft, W.B., "Evaluation of an inference network-based retrieval model", *ACM Transactions on Information Systems*, **9**, 187–222, 1991.

25. van Rijsbergen, C.J., "A new theoretical framework for information retrieval", *Proceedings of the 1986 ACM SIGIR Conference*, ACM, New York, 194–200, 1986.

26. Woods, D.D. and Roth, E.M., *Handbook of Human-Computer Interaction*, Elsevier/North-Holland, Amsterdam, 3–43, 1988.

Chapter 3

Advice from the Oracle: Really Intelligent Information Retrieval

Michael J. Kurtz

Harvard-Smithsonian Center for Astrophysics
60 Garden Street, Cambridge, MA 02138 (USA)
Email: kurtz%cfazwi.decnet@cfa.harvard.edu

3.1 Introduction

What is "intelligent" information retrieval? Essentially this is asking what is intelligence. In this article I will attempt to show some of the aspects of human intelligence, as related to information retrieval. I will do this by the device of a semi-imaginary Oracle. Every Observatory has an oracle, someone who is a distinguished scientist, has great administrative responsibilities, acts as mentor to a number of less senior people, and as trusted advisor to even the most accomplished scientists, and knows essentially everyone in the field.

In an appendix I will present a brief summary of the Statistical Factor Space method for text indexing and retrieval, and indicate how it will be used in the Astrophysics Data System Abstract Service.

3.2 Advice from the Oracle

1. The Oracle sometimes answers without being asked.

Our Oracle walks the hallways, looks into offices, and offers unsolicited advice, often. For example a conversation about the proper definition of galaxy photometry for a particular project was occurring in my office; the Oracle appeared out of nowhere and said "the amplifier on the new chip will not do any better than ten electrons readout

A. Heck and F. Murtagh (eds.), Intelligent Information Retrieval: The Case of Astronomy and Related Space Sciences, 21–28.

noise", then disappeared. This was exactly the key information needed to develop a workable plan.

For machines the equivalent capability will be a long time coming, but there appear to be two central aspects to the problem. (1) How to determine which information is relevant to a problem, and (2) how to determine when relevant information is important enough to state.

To determine relevance the machine might analyze the words in our conversation, and make lists of possibly relevant enough facts. The privacy issues involved in allowing all conversations to pass through a machine which analyzes them will be skirted here.

When the machine should interrupt our conversation to interject some useful information requires rather more thought than just determining which pieces of information might be useful. Certainly having a machine constantly interrupting a conversation with relevant, but not important enough, facts would not be desirable. To be effective the machine must have a detailed knowledge of what the humans know. For example in determining an observing program for galaxy photometry the mean color of galaxies (e.g. $B - R$) is more important than the readout noise of the chip, but both participants in the conversation knew it.

Notice how the privacy problem is exacerbated by the machine storing what each of the participants says, and making a model for the knowledge of each participant.

2. The Oracle knows the environment of the problem area

Q. What photometric system should I use for this CCD study of peculiar A stars in a certain open cluster next December?

O. You should use the Strömgren system, but it is too late in the year. The objects will be too far over in the west, and because of the residual dust from Mt. Pinatubo you will not be able to get the u filter data with small enough errors to constrain your model atmosphere calculations. So wait until next year and do it in September.

Given the question even current systems, such as the ADS Abstract Service, can bring back enough relevant information to give the answer Strömgren Photometry (four of the most relevant 20 articles returned by the ADS contain Strömgren Photometry in their titles). Most of the Oracle's answer could also be gotten by a machine, but not with any existing system. One can easily imagine a rule based system to look up the air mass of a proposed observation, look up the extinction and atmospheric stability at each color, and calculate the expected error. That the atmospheric stability curve is peculiar and due to the volcano would be a note in the database containing the curve.

In order to build a system capable of duplicating the Oracle the data must exist in databases which are connected into a common system, the functions necessary to evaluate the observing conditions would need to exist, and a parser (compiler) would need to translate the question into machine instructions.

With automated literature searches and programs which simulate observations on particular instruments (e.g. on HST), the parts which remain are getting the databases connected, as ADS and ESIS are now doing, and building the rule based system. This rule based system could probably be built with no more effort than has gone into the telescope scheduling program SPIKE (Johnston, 1993). Advances in software methodologies should make this feasible by the end of the decade.

3. The Oracle knows what you need

Q. Should I ask for five or six nights of telescope time for this project?

O. You need to stop taking more data and write up what you've got or your thesis advisor is going to cut your stipend and boot your butt on out of here!

This is not that unusual a question. Everyone has had "Dutch Uncle" conversations like this at some time or another. One might be able to imagine information systems, perhaps first built to aid Time Allocation Committees, which could come up with this answer. What I, anyway, cannot imagine is that I would ever be willing to let a machine talk to me like that, or to accept advice of this nature from a machine. Perhaps I am just old fashioned.

4. The Oracle sometimes answers the question you should have asked

Q. What is the best observing strategy for a redshift survey of Abell 12345?

O. The ST group has about 300 unpublished redshifts in the can already for A12345, better choose another cluster.

This is deceptively simple. It looks like a matter of having connections to the proper databases, and knowing how to search them. Anyone who knew that another group already had 300 redshifts would not begin a cluster study.

How does the machine know this? It cannot be a simple fixed rule, imparted into the program as received wisdom by its creator. Ten years ago 30 redshifts would have been enough to stop any cluster study; now some cluster studies are undertaken despite the existence of more than 100 measurements. One could now get another 300 spectra, for a total of 600, but the bang per buck ratio would be much lower than for choosing another cluster.

The proper rule in the future cannot be predicted now with any certainty, thus we must either employ an expert human regularly updating the rules, or we must have a system complex enough to derive the rules from other data (such as the published literature). Neither is currently feasible. Making the simple conclusion, better choose another cluster. From the simple data, 300 redshifts already exist, will probably be beyond the reasoning abilities of machines for some time.

5. The Oracle can begin a directed dialog to get to an answer

Q. What is the best way to do photometry of spiral galaxies?

O. What do you want to use the photometry for?

Q. I want to get Tully-Fisher distances.

O. Of which galaxies?

Q. From the Zwicky catalog.

O. On the 48" use the Loral 1024 chip with the I filter, 10 minute exposures, and follow the reduction procedures of Bothun and Mould.

The ability to interact with the user is crucial for any information retrieval system, whether man or machine. The trick is asking good questions (or the equivalent). Currently machines can do this by presenting the user with a decision tree of options (e.g.

menu items), by iterating on user chosen "best" answers (e.g. relevance feedback), and other means. This is a key property of intelligent systems.

6. The Oracle is not always right, but always respected.

Q. How should I measure velocity dispersions for elliptical galaxies.

O. Develop a system of isophotal diameters, then measure the spectrum along a long slit over the region defined by the isophotal diameter, correcting for differences in seeing.

This seems like a lot of work. One might imagine that just using a long enough slit and collapsing the spectrum to one dimension using flux weighting (i.e. doing nothing special) could work as well.

The point here is not that the Oracle can be wrong, but that the responsibility for accepting the Oracle's advice rests with the questioner. Since it is all but inconceivable that within our lifetimes a machine will issue scientific judgements which command the trust and respect which the Oracle today enjoys, it follows that very strong limits will exist for some time on what machines will be trusted to do.

7. The Oracle has access to more data than you know exist

Q. I plan to stop off and use the measuring engine for a week after the conference next fall.

O. Ed always takes his vacation then, and he is the only one who can fix that machine if it breaks. It would be safer to go the week before the conference.

Given a vacation schedule to replace the Oracle's educated guess of Ed's vacation plans it is not difficult to imagine a scheduling program giving this answer today. What is missing is the existence and connectivity of the databases and programs. For a machine to match the Oracle a program would have to exist which could access the personnel databases to learn of Ed's work schedule, and could access the measuring engine description, which would have to list Ed as a critical person. This program would have to be accessible to the user, and easy to use.

What the user here is doing is checking an itinerary with the Oracle. Since there are many possible places one might wish to go, and things one might wish to do once one is there, the "simple" task of knowing that Ed might be away can be seen as one of a large number of things which need to be able to be checked. Perhaps if the places one might wish to go all ran scheduling programs which could be accessed by a global scheduling program controlled by the user a capability sufficient to substitute for the Oracle could be created.

8. The Oracle can affect the result, not just derive it

Q. How can I make these observations when we do not have the correct set of filters?

O. If we move some money from the travel budget into the equipment budget we can afford to buy the proper filters. Please write me a one page proposal.

The Oracle is a powerful person, controlling budgets and making decisions about the allocation of resources. While one may cringe at the prospect of a machine cutting the funding for one's trip to a conference, this will certainly happen. The accounting firms are leaders in the field of information systems. Similar decisions are being made now by machine.

9. The Oracle has an opinion in areas of controversy

Q. What is the value of the Hubble constant?
O. $100 kms^{-1} Mpc^{-1}$, by definition.

Intelligent systems will have opinions, this is almost what we mean by intelligent systems.

Feedback loops can help the system's opinions match the user's opinions, but there will always be opinions. What these opinions are will be difficult to control, as they are the end products of complicated interactions of data with algorithms.

10. The Oracle often talks in riddles

Q. The grant has come through, how should I procede?
O. The great prince issues commands, founds states, vests families with fiefs. Inferior people should not be employed.

Whether the advice comes from *I Ching*, our Oracle, or an "intelligent" computer program it is the obligation of the questioner to interpret the information and to decide on the proper action. This remains true, even when the actual "decision" is made by the same machine which provided the information.

To the extent which key decisions are automated, either directly, or by automated decisions as to which information is relevant to bring to the attention of the human decision maker, we lose some control over our lives. For any advice/information system to be beneficial it must empower those decisions which we do make (whether by freeing time which would be wasted making less important decisions, or by providing information, or...) more than it disenfranchises us by making our decisions for us.

3.3 Conclusions

Machines can clearly do some things better than people. Looking through long and boring lists of measurements to select objects with the desired attributes, be they stellar spectra, galaxy images, abstracts of scientific papers, or what have you (Kurtz, 1992), is ideally suited to machines, and is what is normally referred to as "intelligent" information retrieval.

The Oracle does not compete with machines in this way. The Oracle cannot list every low surface brightness galaxy within two degrees of the Coma cluster, nor can the Oracle list every study of Lithium abundances in G stars. Machines can do these things, and will do them with ever increasing frequency.

The Oracle can combine information from a number of different sources to arrive at a conclusion, essentially what would be a vast networked system of interoperating databases exists in the head of the Oracle. Unlike the large data base case, where "intelligent" retrieval is primarily obtained by creating useful similarity measures (metrics), the Oracle sees the logical connection between groups of disparate facts, and comes to a conclusion in a way which has not been previously structured.

We have a long way to go before our machines can retrieve information in a manner as useful as the Oracle. We certainly need to get our databases connected into a common system, so that one database may be examined as a result of the examination of a different database, with these decisions made by a program. We also need to have a mechanism whereby local intelligent processes can communicate with global intelligent processes. Local processes can make use of local expertise, can access local data which could not be made globally available, and can follow local rules in disseminating information. Global processes can examine the results of a number of local processes, combine them with analyses of globally available data and other global processes, and come to conclusions which take all these things into account.

As we progress from the dawn to the first cup of coffee of the Information Age we are faced with the need to make decisions automatically in situations of great complexity. The history of chess playing programs indicates that this will not be a simple endeavor. Machines as intelligent as the Oracle are still just the stuff of science fiction, but as I can now buy a machine for $69 which can beat me at chess every time I can dream that before I die machines will be able to tell me much of what the Oracle can now. By that time the Oracle will have transformed into a Meta-Oracle, giving our progeny a new perspective on the meaning of "intelligence."

Appendix: Statistical Factor Spaces in the Astrophysics Data System

The Astrophysics Data System Abstract Service (Kurtz et al., 1993) is a collaboration af the NASA ADS Project and the NASA Scientific and Technical Information Branch. It provides intelligent access to the NASA-STI abstract databases for astronomy and related fields. Several searching methods have been implemented in this system, each with a number of options. The user has the ability to combine the result of several different simultaneous searches in whichever way seems best. New capabilities are still being added to the system on a regular basis.

Here I will describe the Statistical Factor Space technique, a new search method which is being implemented for the first time in the ADS Abstract Service. It is based on the Factor Space technique of Ossorio (1966), but with the difference that where the original Factor Space relied on human subject matter experts to provide the basic data by filling out extensive questionnaires (essentially a psychological testing type of approach) the Statistical Factor Space obtains the basic data via an *a posteriori* statistical evaluation of a set of classified data (essentially building a psychometric model of the set of librarian classifiers who classified the data).

The basic data element in a Factor Space is a matrix of co-relevancy scores for term (i.e. word or phrase) versus classification (e.g. subject heading or assigned key word) pairs. Thus, for instance the term redshift is highly relevant to the subject heading cosmology, somewhat relevant to the subject heading supernovae, and not relevant to the subject heading stellar photometry. These are the measures which are assigned by humans in the original method. For the NASA STI/ADS system there are more than 10,000,000 of these items; obviously more than is feasible for humans to assign.

A Statistical Factor Space is built by creating the co-occurrence matrix of term–classification pairs, and comparing each element with its expectation value were the

QUERY: 45.066.099 Degeneracies in Parameter Estimates for Models of Gravitational Lens Systems.

1. 45.066.204 Gravitational Lenses.
2. 45.161.354 Light Propagation through Gravitationally Perturbed Friedmann Universes.
3. 45.160.045 Arcs from Gravitational Lensing.
4. 45.066.221 Can We Measure H_0 with VLBI Observations of Gravitational Images?
5. 45.066.012 Gravitational Lensing in a Cold Dark Matter Universe.

Table 3.1: The titles of the query document, and the five closest documents to it in the Factor Space. The numbers are the document numbers assigned by *Astronomy and Astrophysics Abstracts*.

terms distributed at random with respect to the classifications. This comparison cannot be fully parametric, as the expectation value for infrequently appearing terms is less than one, but terms can appear only in integer amounts, so even if they are distributed at random there will be cells in the co-occurrence matrix containing one, a many σ variation.

The comparison function requires the co-occurrence score in a cell to be at least two before it gets weight in the decision. This systematically biases the system against infrequently occurring terms, which is exactly the opposite of the bias of the weighted word matching algorithm available in the ADS system. This suggests that the two techniques (Factor Space and weighted word matches) can be used to complement each other; the ADS system is designed to permit this easily.

The matrix of statistically derived co-relevancy scores is then manipulated using factor analysis to obtain the final relevancy space, where proximity implies nearness of meaning. Documents are then classified as the vector sum of the terms which comprise them, as are queries, and the most relevant documents to a query are simply those with the smallest Euclidean distance to the query in the relevance space. In practice the document vectors are normalized so that they occur on the surface of a unit sphere, so minimizing the Euclidean distance is equivalent to maximizing the cosine of the separation angle.

The ADS Factor Space will be built on the co-occurrence of terms with keywords assigned by the STI librarians, with the dimensionality reduced by factor analysis. A feasibility study has been successfully carried out using this technique with one half year volume of *Astronomy and Astrophysics Abstracts* (vol. 45), there the space was built on the co-occurrence of terms with chapter headings in A&A. Table 3.1 shows an example of how the retrieval works. The query was taken to be an abstract from the volume, number 45.066.099, titled "Degeneracies in Parameter Estimates for Models of Gravitational Lens Systems." In Table 3.1 I show the titles for the closest five papers to that paper, and in Table 3.2 I show the actual terms found in the query, and in each of the closest three documents. Note in particular that the second closest document contains very few words in common, and those are not rare. It is (according to the author of the query document) a very relevant paper. Note that the term "gravitational lens" is used more than 400 times in about 100 different abstracts in the data, most of these papers not as relevant to the

45.066.099	45.066.204	45.161.354	45.160.045
cosmology	black hole	Friedmann universe	background
estimate	catastrophe theory	gravitational lens	clusters of galaxies
gravitational lens	classification	gravitational radiation	computer simulation
image	core	light	core
mass distribution	current	propagation	gravitational lens
mass	galaxy		image
model	gravitational lens		intergalactic medium
observation	law		light
parameter	model		mass distribution
propagation	observation		mass
ray	optics		model
system	position		observation
time	propagation		pair
transform	scaling		symmetry
type	size		
	star		
	topology		

Table 3.2: The terms found in the query document in the left column, and in the three most similar documents from Table 3.1 in the next three columns.

query paper as 45.161.354.

References

1. Kurtz, M.J., "Second Order Knowledge: Information Retrieval in the Terabyte Era", in *Astronomy from Large Databases II*, A. Heck and F. Murtagh, eds., European Southern Observatory, Garching, 85–97, 1992.

2. Kurtz, M.J., Karakashian, T., Stern, C.P., Eichhorn, G., Murray, S.S., Watson, J.M., Ossorio, P.G., and Stoner, J.L., "Intelligent Text Retrieval in the NASA Astrophysics Data System", in *Astronomical Data Analysis and Software and Systems II*, R.J. Hanisch, R.J.V. Brissenden, and J. Barnes, eds., Astronomical Society of the Pacific, San Francisco, 1993, in press.

3. Johnston, M.D., "The SPIKE Scheduling Software" in *Astronomical Data Analysis and Software and Systems II*, R.J. Hanisch, R.J.V. Brissenden, and J. Barnes, eds., Astronomical Society of the Pacific, San Francisco, 1993, in press.

4. Ossorio, P.G., "Classification Space: A Multivariate Procedure for Automatic Document Indexing and Retrieval", *Multivariate Behavioral Research*, 479–524, 1966.

Chapter 4

Search Algorithms for Numeric and Quantitative Data

Fionn Murtagh[1]
Space Telescope – European Coordinating Facility
European Southern Observatory, Karl-Schwarzschild-Str. 2
DW-8046 Garching/Munich (Germany)
Email: fmurtagh@eso.org

'Now', said Rabbit, 'this is a Search, and I've Organised it –'
'Done what to it?' said Pooh.
'Organised it. Which means – well, it's what you do to a Search,
when you don't all look in the same place at once.'
A.A. Milne, The House at Pooh Corner (1928) – M.S. Zakaria

4.1 Efficient Search for the Retrieval of Information

Search algorithms underpin astronomical databases, and may be called upon for the processing of (suitably coded) textual data. They may be required in conjunction with the use of dimensionality reduction approaches such as the factor space approach described in chapter 3, or latent semantic indexing (Deerwester et al., 1990). Efficient search algorithms can be the building blocks of data reorganization approaches using clustering (see section 4.8 below). All in all, search algorithms constitute the motor which drives information retrieval.

This chapter will be primarily concerned with *best match* searching. Other forms of search can be distinguished: for instance, *exact match*, or *partial match*. In the latter cases, either an exact replica of what we have is sought, or a replica of some aspects of what we have, are at issue. The best match, problem, on the other hand confers some initiative on

[1]Affiliated to the Astrophysics Division, Space Science Department, European Space Agency.

A. Heck and F. Murtagh (eds.), Intelligent Information Retrieval: The Case of Astronomy and Related Space Sciences, 29–48.

the retrieval system. A cursory definition of database management systems is that they support exact and partial match retrieval; whereas (a narrow definition of) information retrieval is that it deals with best match retrieval.

When the *fields* or our *records* are numeric, this best match problem is also called the nearest neighbor problem. The record is considered as a point in a multidimensional space, the field values being the projections on the axes of this space. Then the problem is one of finding close points in this space.

The best match or nearest neighbor problem is important in many disciplines. In statistics, k-nearest neighbors, where k can be 1, 2, etc., is a method of non-parametric discriminant analysis. In pattern recognition, this is a widely-used method for unsupervised classification (see Dasarathy, 1991). Nearest neighbor algorithms are also the "building block" of other algorithms, e.g. clustering algorithms; or as effective heuristics for combinatorial optimization algorithms such as the traveling salesman problem, which is a paradigmatic problem in many areas.

In astronomy and related space sciences, data rarely comes in one form only. In fact, we are usually faced with the major forms of: image data, numeric data, and bibliographic data. The topics discussed in this chapter will be most concerned with the latter two areas. In fact, image information retrieval is still mainly carried out through use of ancillary positional or textual data. It would appear that it is mainly with certain categories of graphics data – e.g. engineering drawings in computer aided design, or template structures in geographic information systems – that the situation is other than this (see Grosky and Mehrotra, 1992).

In the numeric data case, the records to be scanned through are usually of fixed length (or, by suitable recoding, are forced to be). The number of fields (assumed atomic or irreducible) or attributes thus gives the dimensionality of each record vector. Many of the algorithms looked at in this chapter are not particularly relevant if the number of records is less than, say, 1000. For such small data-sets, a brute-force checking of each record in order to find the closest is ordinarily quite adequate. In the bibliographic data case, a different problem comes to the fore: namely, that the dimensionality is often very high. In retrieval systems based on keywords or index terms, the latter are often numerous. Hence the number of records, and the number of keywords, may well be comparable. Algorithms looked at later in this chapter indicate certain search strategies in this case.

Especially for the bibliographic case, the reader will find a short overview, which provides background and complementary topics, in Salton (1991) and Kraft (1985). A thorough, and well recommendable, compendium is to be found in Frakes and Baeza-Yates (1992).

4.2 Plan of this Chapter

In the remainder of this chapter, we begin with data structures where the objective is to break the $O(n)$ barrier for determining the nearest neighbor (NN) of a point. Note how a "record" has now become a "point" in a space of dimensionality m, the latter being the associated number of fields or attributes. A brute force scan of all points requires $n-1$ proximity calculations, and following each calculation a check for whether we are closer or not compared to the previous closest neighbor. If we temporarily bypass m, and just consider the number of points n, then the number of calculations is bounded by a constant times n, which we write as "order of n", $O(n)$. The first few algorithms described seek to break this barrier, in the average case if not always in the worst case.

These approaches have been very successful, but they are restricted to low dimensional NN-searching. For higher dimensional data, a wide range of bounding approaches have been proposed, i.e. a provably smallest possible proximity is compared against the current NN proximity. If the former is the greater, then there is no need to proceed to an exact calculation of the proximity. Bounding approaches remain $O(n)$ algorithms. But with a low constant of proportionality such algorithms can be very efficient.

Further sections in this chapter deal with large dimensionality searching, typical of keyword-based systems; and with the use of clustering to expedite search.

The term "proximity" has been used, already, on a number of occasions. One can distinguish between "similarity" on the one hand, or "distance" or "dissimilarity" on the other. Depending on the definition used, it may be feasible to simply convert similarity into dissimilarity by subtraction of the latter from a sufficiently large constant. A distance, like a dissimilarity, satisfies properties of symmetry (i.e. a is as distant from b, as b is from a) and positive definiteness (i.e. necessarily positive or zero; if zero then the points are identical). Additionally, distance satisfies the triangular inequality (geographically expressed: going directly from a to b is always shorter than going via a third point c; unless c lies on the line connecting a and b, in which case we have $d(a,b) = d(a,c) + d(c,b)$). A popular and often used distance is the Euclidean distance:

$$d^2(x, y) = \sum_j (x_{ij} - y_{ij})^2$$

Here the squared distance has been defined: it is often computationally advantageous to use it and dispense with a square root calculation.

Real valued data are most germane to Euclidean distance. It may be necessary to normalize attribute values beforehand, to prevent certain attributes from "shouting louder than others". A basic keyword system may be different: it may involve presence or absence values, coded by 1 and 0, which is referred to as "binary" data. A favored similarity measure has been the so-called Jaccard coefficient:

$$n_{xy}/(n_x + n_y - n_{xy})$$

where n_{xy} is the number of keywords simultaneously present in records x and y; and n_x is the total number of keywords associated with record x.

4.3 Hashing, or Simple Binning or Bucketing

In this approach to NN-searching, a preprocessing stage precedes the searching stage. All points are mapped onto indexed cellular regions of space, so that NNs are found in

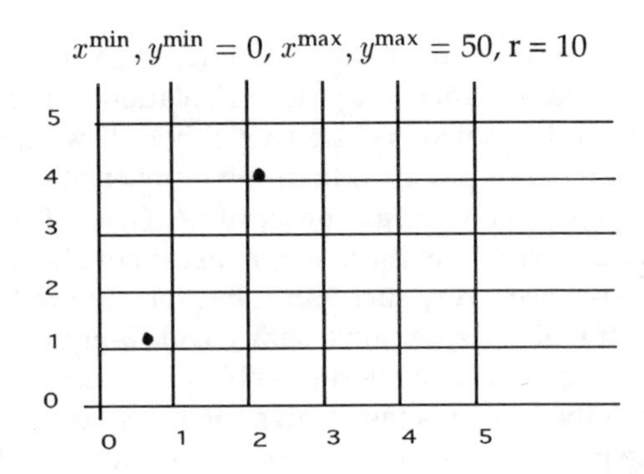

Point (21,40) is mapped onto cell (2,4); point (7,11) is mapped onto cell (0,1).

Figure 4.1: Example of simple binning in the plane.

the same or in closely adjacent cells. Taking the plane as as example, and considering points (x_i, y_i), the maximum and minimum values on all coordinates are obtained (e.g. $(x_j^{\min}, y_j^{\min})$). Consider the mapping (Figure 4.1)

$$x_i \longrightarrow \{\lfloor (x_{ij} - x_j^{\min})/r \rfloor\}$$

where constant r is chosen in terms of the number of equally spaced categories into which the interval $[x_j^{\min}, x_j^{\max}]$ is to be divided. This gives to x_i an integer value between 0 and $\lfloor (x_{ij}^{\max} - x_{ij}^{\min})/r \rfloor$ for each attribute j. $O(nm)$ time is required to obtain the transformation of all n points, and the result may be stored as a linked list with a pointer from each cell identifier to the set of points mapped onto that cell.

NN-searching begins by finding the closest point among those which have been mapped onto the same grid cell as the target point. This gives a current NN point. A closer point may be mapped onto some other grid cell if the distance between target point and current NN point is greater than the distance between the target point and any of the boundaries of the cell containing it. Note that proper corrections will need to be applied so that the grid structure is considered to be superimposed on the original space. If we cannot show that the current NN is necessarily the NN, all grid cells adjacent to the grid cell are searched in order to see if the current NN point can be bettered. It is, of course, possible that the initial cell might have contained no points other than the target point: in this case, the search through adjacent cells may yield a possible NN, and it may be necessary to widen the search to confirm this. Therefore in the plane, 1 cell is searched first; if required, the 8 cells adjacent to this are searched; if necessary, then the up to 16 cells adjacent to these; and so on.

A powerful theoretical result regarding this approach is as follows. For uniformly distributed points, the NN of a point is found in $O(1)$, or constant, expected time (see Delannoy, 1980, or Bentley et al., 1980, for proof). Therefore this approach will work well if approximate uniformity can be assumed or if the data can be broken down into regions of approximately uniformly distributed points.

Simple Fortran code for this approach is listed, and discussed, in Schreiber (1993). The search through adjacent cells requires time which increases exponentially with dimensionality (if it is assumed that the number of points assigned to each cell is approximately equal). As a result, this approach is suitable for low dimensions only. Rohlf (1978) reports on work in dimensions 2, 3, and 4; and Murtagh (1983) in the plane. Rohlf also mentions the use of the first 3 principal components to approximate a set of points in 15-dimensional space. Such dimensionality reduction should be kept in mind in assessing this approach for data of high dimensionality. Linear dimensionality, as provided by principal components analysis, may be bettered, for the purposes of NN-searching, by nonlinear mapping techniques. Examples of the latter are Sammon's nonmetric multidimensional scaling (Sammon, 1969) or Kohonen's topological mapping (Kohonen, 1988).

4.4 Multidimensional Binary Search Tree

A decision tree splits the data to be searched through into two parts: each part is further subdivided; and subdivisions continue until some prespecified number of data points is arrived at. In practice, we associate with each node of the decision tree the definition of a subdivision of the data only, and we associate with each terminal node a pointer to the stored coordinates of the points.

One version of the multidimensional binary search tree (MDBST) is as follows. Halve the set of points, using the median of the first coordinate values of the points. For each of the resulting sets of points, halve them using the median of the second coordinate values. Continue halving; use the medians of all coordinates in succession in order to define successive levels of the hierarchical decomposition; when all coordinates have been exhausted, recycle through the set of coordinates; halt when the number of points associated with the nodes at some level is smaller than or equal to a prespecified constant, c. See example in Figure 4.2. Clearly the tree is kept balanced: this ensures $O(\log n)$ levels, at each of which $O(n)$ processing is required. Hence the construction of the tree takes $O(n \log n)$ time.

The search for a NN then proceeds by a top-down traversal of the tree. The target point is transmitted through successive levels of the tree using the defined separation of the two child nodes at each node. On arrival at a terminal node, all associated points are examined and a current NN selected. The tree is then backtracked: if the points associated with any node could furnish a closer point, then subnodes must be checked out.

The approximately constant number of points associated with terminal nodes (i.e. with hyper-rectangular cells in the space of points) should be greater than 1 in order that some NNs may be obtained without requiring a search of adjacent cells (i.e. of other terminal nodes). Friedman et al. (1977) suggest a value of c between 4 and 32 based on empirical study.

The MDBST approach only works well with small dimensions. To see this, consider each coordinate being used once and once only for the subdivision of points, i.e. each attribute is considered equally useful. Let there be p levels in the tree, i.e. 2^p terminal

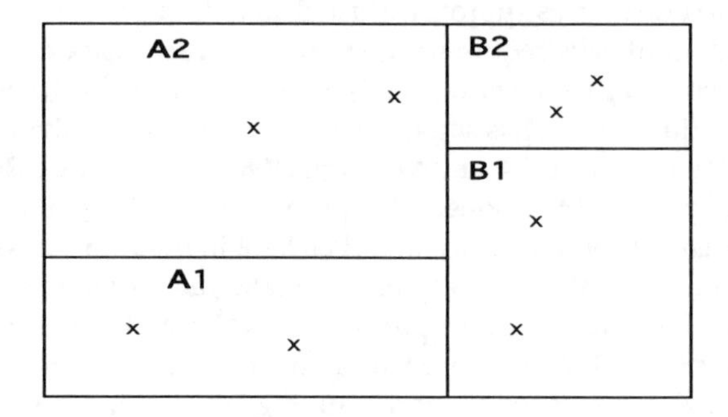

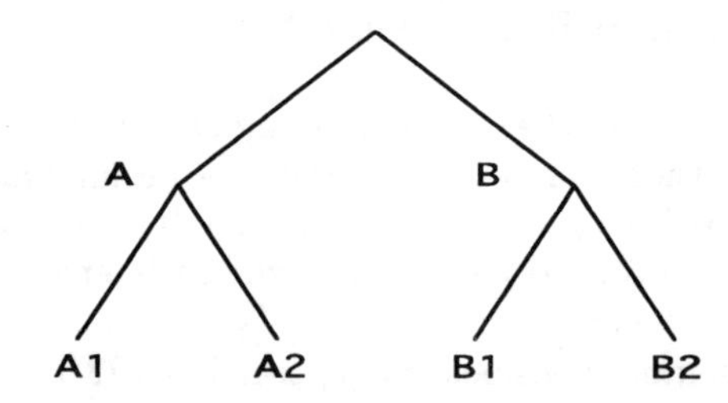

Figure 4.2: A MDBST using planar data.

nodes. Each terminal node contains approximately c points by construction and so $c2^p = n$. Therefore $p = \log_2 n/c$. As sample values, if $n = 32768, c = 32$, then $p = 10$. I.e. in 10-dimensional space, using a large number of points associated with terminal nodes, more than 30000 points will need to be considered. For higher dimensional spaces, two alternative MDBST specifications are as follows.

All attributes need not be considered for splitting the data if it is known that some are of greater interest than others. Linearity present in the data may manifest itself via the variance of projections of points on the coordinates; choosing the coordinate with greatest variance as the discriminator coordinate at each node may therefore allow repeated use of certain attributes. This has the added effect that the hyper-rectangular cells into which the terminal nodes divide the space will be approximately cubical in shape. In this case, Friedman et al. (1977) show that search time is $O(\log n)$ on average for the finding of the NNs of all n points. Results obtained for dimensionalities of between 2 and 8 are reported on in Friedman et al. (1977), and in the application of this approach to minimal spanning tree construction in Bentley and Friedman (1978). Lisp code for the MDBST is discussed in Broder (1990).

The MDBST has also been proposed for very high dimensionality spaces, i.e. where the dimensionality may be greater than the number of points, as could be the case in a keyword-based system. We consider the case of data values indicating association or

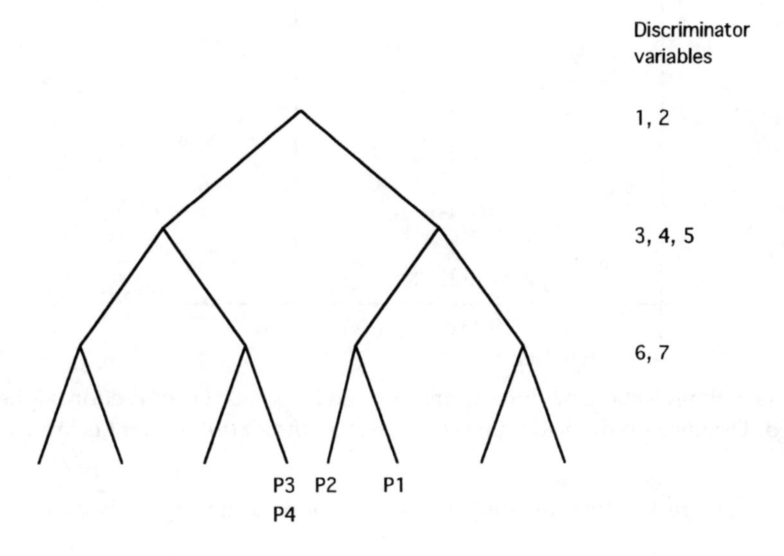

Attributes

		1	2	3	4	5	6	7	8
	p1	1	0	0	0	0	1	1	1
	p2	1	1	0	0	0	0	0	0
Points	p3	0	0	1	0	0	1	1	0
	p4	0	0	0	1	1	1	0	0

Decision rule: presence of some one of discriminating attributes $\Longrightarrow$ take right subtree.

Figure 4.3: MDBST using batching of attributes, given binary data.

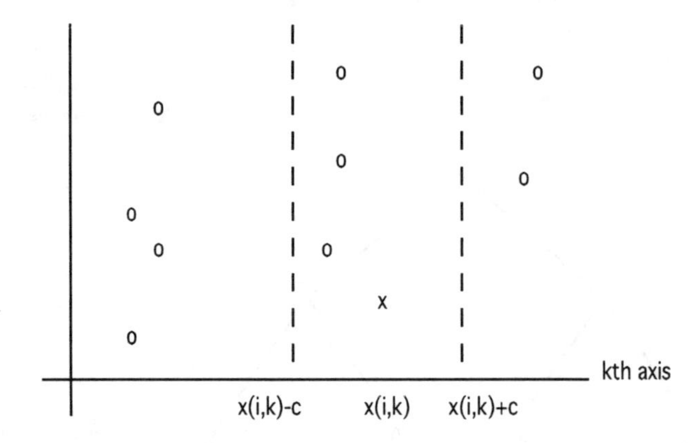

Points with projections within distance c of given point's (x) projection, alone, are searched. Distance c is defined with reference to a candidate or current nearest neighbor.

Figure 4.4: Two-dimensional example of projection-based bound.

non-association (1 or 0) of an index term (attribute) with a record (point). In this context, attributes may be batched together. In Figure 4.3, a target point with 0 coordinate values on attributes 1 and 2 is directed towards the left child node of the root; otherwise it is sent to the right child node. At level 2, if the target point has 0 coordinates for attributes 3, 4, and 5, it is directed towards the left subtree and otherwise towards the right subtree. As with the version of the MDBST described above, a top-down traversal of the tree leads to the search of all points associated with a terminal node. This provides a current NN. Then backtracking is commenced in order to see if the current NN can be bettered. Large n, well in excess of 1400, was stated as necessary for good results (Weiss, 1981; Eastman and Weis, 1982). General guidelines for the attributes which define the direction of search at each level are that they be related, and the number chosen should keep the tree balanced.

4.5 Bounding using Projections or Properties of Metrics

Making use of bounds is a versatile approach, which may be less restricted by dimensionality. Some lower bound on the dissimilarity is efficiently calculated in order to dispense with the full calculation in many instances.

Using projections on a coordinate axis allows the exclusion of points in the search for the NN of point x_i. Points x_k, only, are considered such that $(x_{ij} - x_{kj})^2 \leq c^2$ where x_{ij} is the jth coordinate of x_i, and where c is some prespecified distance (see Figure 4.4).

Alternatively more than one coordinate may be used. The prior sorting of coordinate values on the chosen axis or axes expedites the finding of points whose full distance calculation is necessitated. The preprocessing required with this approach involves the sorting of up to m sets of coordinates, i.e. $O(mn \log n)$ time.

Using one axis, it is evident that many points may be excluded if the dimensionality is very small, but that the approach will disimprove as the latter grows. Friedman et al. (1975) give the expected NN search time, under the assumption that the points are uniformly distributed, as $O(mn^{1-1/n})$. This approaches the brute force $O(nm)$ as m gets large. Reported empirical results are for dimensions 2 to 8.

Marimont and Shapiro (1979) extend this approach by the use of projections in subspaces of dimension greater than 1 (usually about $m/2$ is suggested). This can be further improved if the subspace of the principal components is used. Dimensions up to 40 are examined.

The Euclidean distance is very widely used. Two other members of a family of metric measures – the family is referred to as the Minkowski metrics – require less computation time to calculate, and they can be used to provide bounds on the Euclidean distance. We have:

$$d_1(x, x') \geq d_2(x, x') \geq d_\infty(x, x')$$

where d_1 is the Hamming distance defined as $\sum_j \mid x_j - x'_j \mid$; the Euclidean distance is given by the square root of $\sum_j (x_j - x'_j)^2$; and the Chebyshev distance is defined as $\max_j \mid x_j - x'_j \mid$.

Kittler (1978) makes use of the following bounding strategy: *reject all points y such that* $d_1(x, y) \geq \sqrt(m)\delta$ where δ is the current NN d_2-distance. The more efficiently calculated d_1-distance may thus allow the rejection of many points (90% in 10-dimensional space is reported on by Kittler). Kittler's rule is obtained by noting that the greatest d_1-distance between x and x' is attained when

$$\mid x_j - x'_j \mid^2 = d_2^2(x, x')/m$$

for all coordinates, j. Hence $d_1(x, x') = d_2(x, x')/\sqrt{m}$ is the greatest d_1-distance between x and x'. In the case of the rejection of point y, we then have:

$$d_1(x, y) \leq d_2(x, y)/\sqrt{m}$$

and since, by virtue of the rejection,

$$d_1(x, y) \geq \sqrt{m}\delta$$

it follows that $\delta \leq d_2(x, y)$.

Yunck (1976) presents a theoretical analysis for the similar use of the Chebyshev metric. Richetin et al. (1980) propose the use of both bounds. Using uniformly distributed points in dimensions 2 to 5, the latter reference reports the best outcome when the rule: *reject all y such that $d_\infty(x, y) \geq \delta$* precedes the rule based on the d_1-distance. Up to 80% reduction in CPU time is reported.

4.6 Bounding using the Triangular Inequality

The triangular inequality is satisfied by distances: $d(x, y) \leq d(x, z) + d(z, y)$, where x, y and z are any three points. The use of a *reference point*, z, allows a full distance calculation between point x, whose NN is sought, and y to be avoided if

$$| d(y, z) - d(x, z) | \geq \delta$$

where δ is the current NN distance. The set of all distances to the reference point are calculated and stored in a preprocessing step requiring $O(n)$ time and $O(n)$ space. The above cut-off rule is obtained by noting that if

$$d(x, y) \geq | d(y, z) - d(x, z) |$$

then, necessarily, $d(x, y) \geq \delta$. The former inequality above reduces to the triangular inequality irrespective of which of $d(y, z)$ or $d(x, z)$ is the greater.

The set of distances to the reference point, $\{d(x, z) \mid x\}$, may be sorted in the preprocessing stage. Since $d(x, z)$ is fixed during the search for the NN of x, it follows that the cut-off rule will not then need to be applied in all cases.

The single reference point approach, due to Burkhard and Keller (1973), was generalized to multiple reference points by Shapiro (1977). The sorted list of distances to the first reference point, $\{d(x, z_1) \mid x\}$, is used as described above as a preliminary bound. Then the subsequent bounds are similarly employed to further reduce the points requiring a full distance calculation. The number and the choice of reference points to be used is dependent on the distributional characteristics of the data. Shapiro (1977) finds that reference points ought to be located away from groups of points. In 10-dimensional simulations, it was found that at best only 20% of full distance calculations were required (although this was very dependent on the choice of reference points).

Hodgson (1988) proposes the following bound, related to the training set of points, y, among which the NN of point x is sought. Determine in advance the NNs and their distances, $d(y, NN(y))$ for all points in the training set. For point y, then consider $\delta_y = \frac{1}{2} d(y, NN(y))$. In seeking NN($x$), and having at some time in the processing a candidate NN, y', we can exclude all y from consideration if we find that $d(x, y') \leq \delta_{y'}$. In this case, we know that we are sufficiently close to y' that we cannot improve on it.

We return now to the choice of reference points: Vidal Ruiz (1986) proposes the storing of inter-point distances between the members of the training set. Given x, whose NN we require, some member of the training set is used as a reference point. Using the bounding approach based on the triangular inequality, described above, allows other training set members to be excluded from any possibility of being NN(x). Micó et al. (1992) and Ramasubramanian and Paliwal (1992) discuss further enhancements to this approach, focused especially on the storage requirements.

Fukunaga and Narendra (1975) make use of both a hierarchical decomposition of the data set (they employ repeatedly the k-means partitioning technique), and bounds based on the triangular inequality. For each node in the decomposition tree, the center and maximum distance to the center of associated points (the "radius") are determined. For 1000 points, 3 levels were used, with a division into 3 classes at each node.

All points associated with a non-terminal node can be rejected in the search for the NN of point x if the following rule (Rule 1) is not verified:

$$d(x, g) - r_g < \delta$$

where δ is the current NN distance, g is the center of the cluster of points associated with the node, and r_g is the radius of this cluster. For a terminal node, which cannot be rejected on the basis of this rule, each associated point, y, can be tested for rejection using the following rule (Rule 2):

$$| d(x,g) - d(y,g) | \geq \delta.$$

These two rules are direct consequences of the triangular inequality.

A branch and bound algorithm can be implemented using these two rules. This involves determining some current NN (the bound) and subsequently branching out of a traversal path whenever the current NN cannot be bettered. Not being inherently limited by dimensionality, this approach appears particularly attractive for general purpose applications.

Other rejection rules are considered by Kamgar-Parsi and Kanal (1985). A simpler form of clustering is used in the variant of this algorithm proposed by Niemann and Goppert (1988). A shallow MDBST is used, followed by a variant on the branching and bounding described above.

4.7 High-Dimensional Sparse Binary Data

"High-dimensional", "sparse" and "binary" are the characteristics of keyword-based bibliographic data. Values in excess of 10000 for both n and m would not be unusual. The particular nature of such data – including very sparse binary document-term associations – has given rise to particular NN-searching algorithms. Such data is usually stored as compact lists, i.e. the sequence number of the document followed by the sequence number of the terms associated with it. Commercial document collections are usually searched using a Boolean search environment: i.e. all documents associated with particular terms are retrieved; the intersection/union/etc. of such sets of documents are obtained using AND, OR and other connectives. For efficiency, an *inverted file* which maps terms onto documents is available for Boolean retrieval. An inverted file is stored in a similar manner to the document-term file. The efficient NN algorithms, to be discussed, make use of both the document-term and the term-document files.

The usual algorithm for NN-searching considers each document in turn, calculates the distance with the given document, and updates the NN if appropriate. This algorithm is shown schematically in Figure 4.5 (top). The inner loop is simply an expression of the fact that the distance or similarity will, in general, require $O(m)$ calculation: examples of commonly used coefficients are the Jaccard similarity, and the Hamming distance (both mentioned earlier in this chapter).

If $\bar{m}$ and $\bar{n}$ are, respectively, the average numbers of terms associated with a document, and the average number of documents associated with a term, then an average complexity measure, over n searches, of this usual algorithm is $O(\bar{m}n)$. It is assumed that advantage is taken of some packed form of storage in the inner loop (e.g. using linked lists).

Croft's algorithm (see Croft, 1977, and Figure 4.5) is of worst case complexity $O(nm^2)$. However the number of terms associated with the document whose NN is required will often be quite small. The National Physical Laboratory test collection, for example, which

```
Usual algorithm:

       Initialize current NN
       For all documents in turn do:
       ... For all terms associated with the document do:
       ... ... Determine (dis)similarity
       ... Endfor
       ... Test against current NN
       Endfor


Croft's algorithm:

       Initialize current NN
       For all terms associated with the given document do:
       ... For all documents associated with each term do:
       ... ... For all terms associated with a document do:
       ... ... ... Determine (dis)similarity
       ... ... Endfor
       ... ... Test against current NN
       ... Endfor
       Endfor


Perry-Willett algorithm:

       Initialize current NN
       For all terms associated with the given document, i, do:
       ... For all documents, i', associated with each term, do:
       ... ... Increment location i' of counter vector
       ... Endfor
       Endfor
```

Figure 4.5: Algorithms for NN-searching using high-dimensional sparse binary data.

was used in Murtagh (1982) has the following characteristics: n = 11429, m = 7491, $\bar{m}$ = 19.9, and $\bar{n}$ = 30.4. The outermost and innermost loops in Croft's algorithm use the document-term file. The center loop uses the term-document inverted file. An average complexity measure (more strictly, the time taken for best match search based on an average document with associated average terms) is seen to be $O(\bar{n}\bar{m}^2)$.

In the outermost loop of Croft's algorithm, there will eventually come about a situation where – if a document has not been thus far examined – the number of terms remaining for the given document do not permit the current NN document to be bettered. In this case, we can cut short the iterations of the outermost loop. The calculation of a bound, using the greatest possible number of terms which could be shared with a so-far unexamined document has been exploited by Smeaton and van Rijsbergen (1981) and by Murtagh (1982) in successive improvements on Croft's algorithm.

The complexity of all the above algorithms has been measured in terms of operations to be performed. In practice, however, the actual accessing of term or document information may be of far greater cost. The document-term and term-document files are ordinarily stored on direct access file storage because of their large sizes. The strategy used in Croft's algorithm, and in improvements on it, does not allow any viable approaches to batching together the records which are to be read successively, in order to improve accessing-related performance.

The Perry-Willett algorithm (see Perry and Willett, 1983) presents a simple but effective solution to the problem of costly I/O. It focuses on the calculation of the number of terms common to the given document x and each other document, y in the document collection. This set of values is built up in a computationally efficient fashion. $O(n)$ operations are subsequently required to determine the (dis)similarity, using another vector comprising the total numbers of terms associated with each document. Computation time (the same "average" measure as that used above) is $O(\bar{n}\bar{m} + n)$. We now turn attention to numbers of direct-access reads required.

In Croft's algorithm, all terms associated with the document whose NN is desired may be read in one read operation. Subsequently, we require $\bar{n}\bar{m}$ reads, giving in all $1 + \bar{n}\bar{m}$. In the Perry-Willett algorithm, the outer loop again pertains to the one (given) document, and so all terms associated with this document can be read and stored. Subsequently, $\bar{m}$ reads, i.e. the average number of terms, each of which demands a read of a set of documents, are required. This gives, in all, $1 + \bar{m}$. Since these reads are very much the costliest operation in practice, the Perry-Willett algorithm can be recommended for large values of n and m. Its general characteristics are that (i) it requires, as do all the algorithms discussed in this section, the availability of the inverted term-document file; and (ii) it requires in-memory storage of two vectors containing n integer values.

4.8 Cluster-Based Retrieval

Sets of nearest neighbors constitute clusters, and as will be seen below, then can also be used as building blocks for widely-used hierarchical or agglomerative methods. So far in this chapter we have been concerned with search. Clustering has been used for quite varied objectives in bibliographic retrieval systems, viz.:

- Keyword clustering has been mentioned in section 4.4, above, as a way to improve the performance of a tree-structuring of the training set. More generally, a keyword clustering might aim at automating the task of building synonym lists. Such clusters of keywords can be used to "complete" query terms input by a user, in order to make a stronger and more effective augmented query.

- Co-citations, i.e. sets of documents which cite to a greater or lesser extent the same documents, give an alternative way to define clusters of documents. These clusters can be used to identify fields of research, "and in this way to reveal a structure based on the consensual publishing and referencing practices" of particular scientific research communities (Kimberley, 1985).

- Browsing and current-awareness usages of bibliographic information retrieval systems implicitly or explicitly use clustering. Retrieval, here, is part of the picture: it is complemented by navigation. It may be necessary to support a "stream of consciousness" type of navigation, or to seek "islands of similarity" or clusters.

- And in best match retrieval, a more sophisticated structuring of the data may be required, over and above some of the techniques described earlier in this chapter. This may be based on a set of non-overlapping clusters, such that a "bag of monkeys" retrieval strategy can be used (cf. the arcade vending machine in which a retrieved "monkey" holds onto another, and the latter onto another). A single-pass, and hence very efficient, algorithm for such a purpose is described in Salton and McGill (1983). A top-down scan strategy has been investigated by Griffiths et al. (1984). This uses some hierarchical clustering strategy as a preprocessing stage, and retrieval processing is carried out in a similar vein to what was described earlier in this chapter (section 4.4).

The very close link between nearest neighbor finding, and clustering, is buttressed by the fact that many important agglomerative algorithms can be efficiently implemented by searching for reciprocal nearest neighbors: these are points x and y such that $y = NN(x)$ and $x = NN(y)$. In the case of most well-known agglomerative criteria, reciprocal nearest neighbors can be considered, immediately, as clusters. An allied graph-theoretic structure which is of use is the NN-chain: this is a series such that the second member is the NN of the first, the third member is the NN of the second, and so on until the chain necessarily ends in a reciprocal nearest neighbor pair. Different agglomerative criteria are specified in accordance with how they define the cluster which replaces a reciprocal nearest neighbor pair. Encompassed are, at one extreme, linkage-based clustering, which can be expressed in terms of connected components and includes such structures as the minimal spanning tree; and synoptic clustering, based on graph-theoretic cliques, or minimal variance. For more details of these algorithms, see Murtagh (1985) and Murtagh and Heck (1987).

4.9 Range Searching, The Quadtree and the Sphere Quadtree

A wealth of search problems involving 2- or 3-dimensional data have been studied in the area of computational geometry. Apart from location-oriented search, other objectives include the construction of particular data structures: convex hull, Voronoi diagram, triangulations, etc.; intersection problems involving lines and polygons; and enclosure and overlap problems (e.g. largest empty rectangle, smallest enclosing triangle, etc.). For an overview of such algorithms, see Preparata and Shamos (1985).

Apart from location search problems, another class of search problems involve range searching. One wishes either to retrieve, or to count, the number of points (x, y) satisfying $a \leq x \leq b$ and $c \leq y \leq d$ where the parameters (a, b, c, d) are user-specified. Clearly, one can examine each point in turn and see if it respects the specified constraints. This gives an $O(n)$ algorithm, if we are dealing with n points in all. The prime objective of efficient algorithms is to get this computational requirement down to $O(\log n)$ (motivated by our ability to search for a given number among n numbers using a binary search tree, not to mention various other results already discussed in this chapter). The use of the multidimensional binary search tree is one of the methods discussed by Bentley and Friedman (1979) and Preparata and Shamos (1985) for tackling this problem.

In computational geometry, interrelationships between points and other objects are algorithmically studied. In image processing and in geographically referenced data, the basic objects are often given in an a priori regular relationship, in pixelated regions. With such raster data, one commonly-used hierarchical data structure is the quadtree in the 2-dimensional image case (Figure 4.6), or the octree in the case of a 3-dimensional data cube. The quadtree partitions a gray-scale image at first into 4 quadrants. Let these be called N, S, E and W. Each of these quadrants are recursively partitioned, if they do not satisfy a homogeneity criterion. In the case of a gray-level image with a fixed (small) number of discrete gray levels, the homogeneity criterion might be that all pixels have the same value. Hence terminal or leaf nodes of the quadtree are associated with irreducible image regions. Large, homogeneous image regions yield shallow subtrees, while "busy" areas cause deep subtrees to be created.

Nearest neighbor searching using the quadtree – e.g. finding the pixel or pixels of closest color to a given pixel – is carried out in a similar manner to the top-down recursive search functionality supported by the multidimensional binary search tree (section 4.4 above). Samet (1984) and Samet and Webber (1988) discuss the use of the quadtree for various other operations, – e.g. set-theoretic operations on images; image rotation and other transformations; calculating areas and moments; connected component labeling; etc.

For point or region data, the quadtree suffers when such data is inherently spherical. Planar representations of spherical data have familiar distortions at the poles. To generate planar views of neighborhoods, cartographic projections must be brought into play. Square regions are extremely blunt tools to use when one is seeking sparse or dense areas. An alternative to the quadtree for such data is the *sphere quadtree*.

Fekete (1990) considers this generalization of the quadtree for data on the sphere. The sphere is first approximated by a convex polyhedron. A natural choice for this is

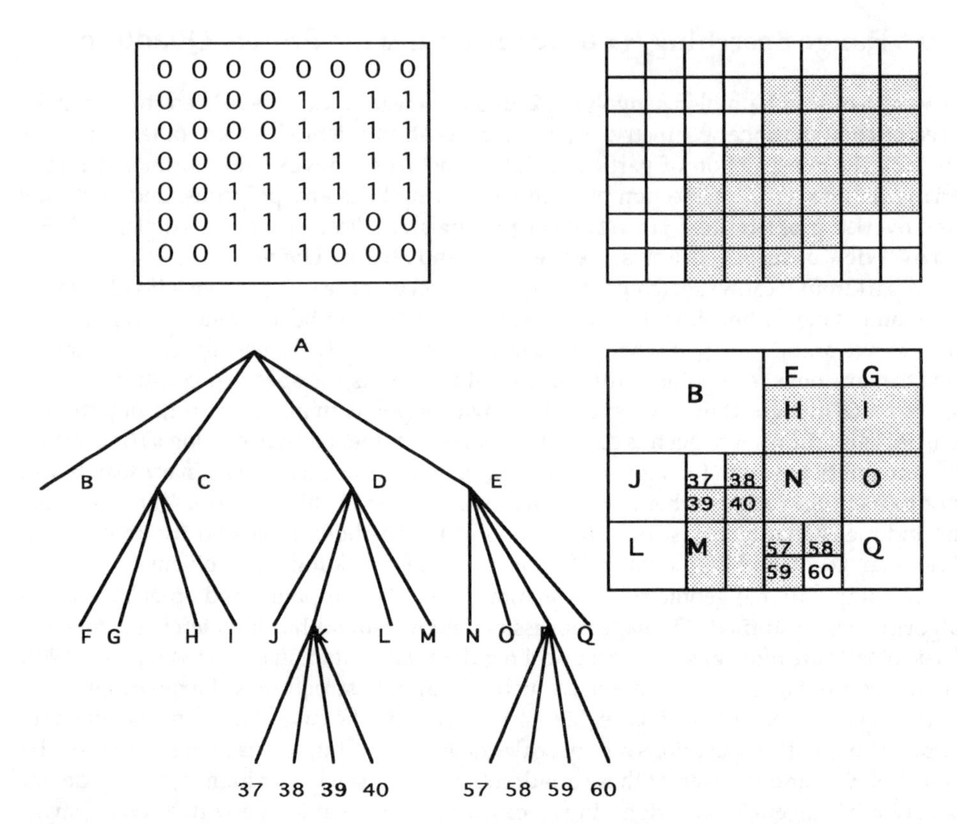

Figure 4.6: A quadtree of region data. For point data, the nodes can be defined by a density criterion. Note how the quadtree generalizes the multidimensional binary search tree (cf. section 4.4), since it is based on 4-way node splitting rather than 2-way.

based on a regular triangular tesselation, yielding an icosahedron. A greater number of faces gives a finer mesh, and a more accurate representation. Thus, the coarse tiling is recursively repeated to any desired level of depth in the sphere quadtree (cf. Figure 4.7). The resulting primitive image regions, associated with nodes of the sphere quadtree, are termed *trixels*.

Operations using the *sphere quadtree* are then addressed. A simple labeling scheme can be used to facilitate checking whether or not two trixels are adjacent. Going on from this, a basic operation needed for many applications is to determine all neighboring trixels to a given trixel.

The sphere quadtree data structure has been used for remotely sensed earth data, and has been mooted for use with the Hubble Space Telescope Data Archive and Distribution System (DADS) as part of the user interface for data selection.

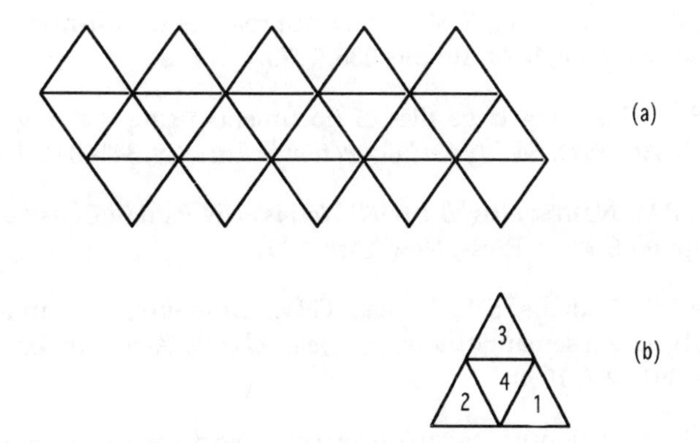

Figure 4.7: (a) The top level (necessarily 20) trixels of a sphere quadtree, representing a sphere; (b) Children of a trixel.

4.10 Other Issues Arising

We have dealt with a number of approaches to supporting search in the case of *static* data collections, with an emphasis on numerical data or data which can be quantitatively coded. *Dynamic* data sets, involving insertions and deletions, can be supported by some of these approaches, but often with very substantial overhead costs.

We have also dealt with approaches which are immediately compatible with the dominant Von Neumann computational paradigm. Some of the wide-ranging research activities in this area, which are based on parallel approaches, are surveyed in a special issue of *Information Processing and Management*, vol. 27, 1991.

Finally, we have not dealt with the important enhancements to the search objectives which are concomitant with fuzziness: Subramanian et al. (1986) or Miyamoto (1990) may be referred to for discussion on this topic.

References

1. Bentley, J.L. and Friedman, J.H., "Fast algorithms for constructing minimal spanning trees in coordinate spaces", *IEEE Transactions on Computers*, **C-27**, 97–105, 1978.

2. Bentley, J.L. and Friedman, J.H., "Data structures for range searching", *ACM Computing Surveys*, **11**, 397–409, 1979.

3. Bentley, J.L., Weide, B.W. and Yao, A.C., "Optimal expected time algorithms for closest point problems", *ACM Transactions on Mathematical Software*, **6**, 563–580, 1980.

4. Broder, A.J., "Strategies for efficient incremental nearest neighbor search", *Pattern Recognition*, **23**, 171–178, 1990.

5. Burkhard, W.A. and Keller, R.M., "Some approaches to best-match file searching", *Communications of the ACM*, **16**, 230–236, 1973.

6. Croft, W.B., "Clustering large files of documents using the single-link method", *Journal of the American Society for Information Science*, **28**, 341–344, 1977.

7. Dasarathy, B.V., *Nearest Neighbor (NN) Norms: NN Pattern Classification Techniques*, IEEE Computer Society Press, New York, 1991.

8. Deerwester, S., Dumais, S.T., Furnas, G.W., Landauer, T.K. and Harshman, R., "Indexing by latent semantic indexing", *Journal of the American Society of Information Science*, **41**, 391–407, 1990.

9. Delannoy, C., "Un algorithme rapide de recherche de plus proches voisins", *RAIRO Informatique/Computer Science*, **14**, 275–286, 1980.

10. Eastman, C.M. and Weiss, S.F., "Tree structures for high dimensionality nearest neighbor searching", *Information Systems*, **7**, 115–122, 1982.

11. Fekete, G., "Rendering and managing spherical data with sphere quadtrees", *Proceedings of the First IEEE Conference on Visualization*, IEEE Computer Society, Los Alamitos, CA, 1990.

12. Frakes, W.B. and Baeza-Yates, R., eds., *Information Retrieval: Data Structures and Algorithms*, Prentice-Hall, Englewood Cliffs, NJ, 1992.

13. Friedman, J.H., Baskett, F. and Shustek, L.J., "An algorithm for finding nearest neighbors", *IEEE Transactions on Computers*, **C-24**, 1000-1006, 1975.

14. Friedman, J.H., Bentley, J.L. and Finkel, R.A., "An algorithm for finding best matches in logarithmic expected time", *ACM Transactions on Mathematical Software*, **3**, 209–226, 1977.

15. Fukunaga, K. and Narendra, P.M., "A branch and bound algorithm for computing k-nearest neighbors", *IEEE Transactions on Computers*, **C-24**, 750–753, 1975.

16. Griffiths, A., Robinson, L.A. and Willett, P., "Hierarchic agglomerative clustering methods for automatic document classification", *Journal of Documentation*, **40**, 175–205, 1984.

17. Grosky, W.I. and Mehrotra, R., "Image database management", in M.C. Yovits, ed., *Advances in Computers Vol. 35*, Academic Press, New York, 237–291, 1992.

18. Hodgson, M.E., "Reducing the computational requirements of the minimum-distance classifier", *Remote Sensing of Environment*, **25**, 117–128, 1988.

19. Kamgar-Parsi, B. and Kanal, L.N., "An improved branch and bound algorithm for computing k-nearest neighbors", *Pattern Recognition Letters*, **3**, 7–12, 1985.

20. Kimberley, R., "Use of literature-based model to define research activities: Applications in research planning", Institute for Scientific Information (ISI) paper, 14 pp., 1985.

21. Kittler, J., "A method for determining k-nearest neighbors", *Kybernetes*, 7, 313–315, 1978.

22. Kohonen, T., "Self-organizing topological maps (with 5 appendices)", in *Proc. Connectionism in Perspective: Tutorial*, University of Zurich, 1988.

23. Kraft, D.H., "Advances in information retrieval: where is that /#*&@ record?", in M.C. Yovits, Ed., *Advances in Computers Vol. 24*, 277–318, 1985.

24. Micó, L., Oncina, J. and Vidal, E., "An algorithm for finding nearest neighbors in constant average time with a linear space complexity", in *11th International Conference on Pattern Recognition*, Volume II, IEEE Computer Science Press, New York, 557–560, 1992.

25. Marimont, R.B. and Shapiro, M.B., "Nearest neighbor searches and the curse of dimensionality", *Journal of the Institute of Mathematics and Its Applications*, **24**, 59–70, 1979.

26. Miyamoto, S., *Fuzzy Sets in Information Retrieval and Cluster Analysis*, Kluwer, Dordrecht, 1990.

27. Murtagh, F., "A very fast, exact nearest neighbor algorithm for use in information retrieval", *Information Technology*, **1**, 275–283, 1982.

28. Murtagh, F., "Expected time complexity results for hierarchic clustering algorithms which use cluster centers", *Information Processing Letters*, **16**, 237–241, 1983.

29. Murtagh, F., *Multidimensional Clustering Algorithms*, Physica-Verlag, Würzburg, 1985.

30. Murtagh, F. and Heck, A., *Multivariate Data Analysis*, Kluwer Academic, Dordrecht, 1987.

31. Niemann, H. and Goppert, R., "An efficient branch-and-bound nearest neighbor classifier", *Pattern Recognition Letters*, **7**, 67–72, 1988.

32. Perry, S.A. and Willett, P., "A review of the use of inverted files for best match searching in information retrieval systems", *Journal of Information Science*, **6**, 59–66, 1983.

33. Preparata, F.P. and Shamos, M.I., *Computational Geometry*, Springer-Verlag, New York, 1985.

34. Ramasubramanian, V. and Paliwal, K.K., "An efficient approximation-algorithm for fast nearest-neighbor search based on a spherical distance coordinate formulation", *Pattern Recognition Letters*, **13**, 471–480, 1992.

35. Richetin, M., Rives, G. and Naranjo, M.,"Algorithme rapide pour la détérmination des k plus proches voisins", *RAIRO Informatique/Computer Science*, **14**, 369–378, 1980.

36. Rohlf, F.J., "A probabilistic minimum spanning tree algorithm", *Information Processing Letters*, **7**, 44–48, 1978.

37. Salton, G., "Developments in automatic text retrieval", *Science*, **253**, 974–1012, 1991.

38. Salton, G. and McGill, M.J., *Introduction to Modern Information Retrieval*, McGraw-Hill, New York, 1983.

39. Samet, H., "The quadtree and related hierarchical data structures", *ACM Computing Surveys*, **16**, 187–260, 1984.

40. Samet, H. and Webber, R.E., "Hierarchical data structures and algorithms for computer graphics", *IEEE Computer Graphics and Applications*, May, 48–68, 1988.

41. Sammon, J.W., "A nonlinear mapping for data structure analysis", *IEEE Transactions on Computers*, **C-18**, 401–409, 1969

42. Schreiber, T., "Efficient search for nearest neighbors", in A.S. Weigend and N.A Gershenfeld, eds., *Predicting the Future and Understanding the Past: A Comparison of Approaches*, Addison-Wesley, New York, in press, 1993.

43. Shapiro, M., "The choice of reference points in best-match file searching", *Communications of the ACM*, **20**, 339–343, 1977.

44. Smeaton, A.F. and van Rijsbergen, C.J., "The nearest neighbor problem in information retrieval: an algorithm using upperbounds", *ACM SIGIR Forum*, **16**, 83–87, 1981.

45. Subramanian, V., Biswas, G. and Bezdek, J.C., "Document retrieval using a fuzzy knowledge-based system", *Optical Engineering*, **25**, 445–455, 1986.

46. Vidal Ruiz, E., "An algorithm for finding nearest neighbors in (approximately) constant average time", *Pattern Recognition Letters*, **4**, 145–157, 1986.

47. Weiss, S.F., "A probabilistic algorithm for nearest neighbor searching", in R.N. Oddy et al., Eds., *Information Retrieval Research*, Butterworths, London, 325–333, 1981.

48. Yunck, T.P., "A technique to identify nearest neighbors", *IEEE Transactions on Systems, Man, and Cybernetics*, **SMC-6**, 678–683, 1976.

Information-Sifting Front Ends to Databases

Hans-Martin Adorf

Space Telescope – European Coordinating Facility
European Southern Observatory, Karl-Schwarzschild-Str. 2
DW-8046 Garching b. München (Germany)
Email: adorf@eso.org, eso::adorf

Two years of research
can easily save you
a week in the library.
– Larry November (1992)

4.1 Introduction

The acquisition, recording, organization, retrieval, display and dissemination of large quantities of data have become standard practice in modern science including astronomy. The amount of knowledge generated within the span of a scientist's career together with the ubiquitous high degree of professional specialization generates problems of its own. Knowledge revealed in one discipline has little time to naturally diffuse into neighbouring disciplines where it might become beneficial, sometimes even be urgently needed.

Also within a single scientific discipline information dissemination problems prevail. Professional astronomical journals have been witnessing an exponential growth in publication rate. For instance, the number of articles related to galaxies – admittedly one of the fastest growing sectors – has been estimated to be doubling every eight years (see Helou et al., 1991).

Discoveries are being made repeatedly. A case in point is the iterative maximum likelihood restoration algorithm for Poisson-noise data, which has been independently discovered (at least) four times within 14 years (Tarasko, 1969; Richardson, 1972; Lucy, 1974; Shepp and Vardi, 1982). Another example, taken from uncertainty reasoning, is the discovery that there is essentially only a single method for consistently combining

A. Heck and F. Murtagh (eds.), Intelligent Information Retrieval: The Case of Astronomy and Related Space Sciences, 49–80.

independent evidence, a theorem independently found at least three times (Cox, 1946; Hájek, 1985; Cheng and Kashyap, 1988).

Among the reasons responsible for this duplication of effort one finds that, while the amount of information has grown very fast, the sophistication of the tools for accessing the information "mountain" has not evolved at comparable pace. The search for and retrieval of information often is no fun and, when carried out by traditional means, also costly. Thus scientific libraries, the traditional repositories for organized knowledge, often act as information sinks, rather than inspiring sources.

Fortunately, much of today's information is already stored on digital media. Thus it can be processed by computers and it can, in principle, be easily disseminated via the flourishing computer networks (Carpenter and Williams, 1992; Hanisch, 1992a, b; Krol, 1992; Press, 1992a, b). Many astronomical databases, which traditionally have been available only in print, were later transferred to magnetic tape, and are now on-line accessible, at least in principle, either locally or on the wide-area computer network (Murtagh, 1990; Albrecht and Egret, 1991, 1992).

This welcome development also entails problems: the *resource discovery* problem (What exists? Where is it? How to access it?); the problem of human *cognitive overload* and that of *electronic junk* (Denning, 1982). Using computers for information retrieval and information filtering (Belkin and Croft, 1992) is becoming a necessity.

Below I shall investigate several database management systems (DBMSs) and note features which facilitate the process of sifting through large amounts of scientific information. Unfortunately, to my knowledge, no synopsis has been published explaining the access mechanisms to different astronomical databases and assessing their capabilities in an objective and homogeneous way. Such a comparison, based on actual experience with the individual astronomical databases, is certainly highly desirable, but beyond the scope of this contribution. Instead, I have to confine myself mainly to published material on astronomical databases. I shall describe a few commercial/public domain DBMSs and use the latter to draw some conclusions about the likely future evolution of astronomical databases, particularly with respect to their user interface (UIF) functionality.

4.2 Astronomical Databases – Some Characteristics

Astronomical databases belong to the class of scientific and statistical databases (SSDBMS, Shoshani and Wong, 1985; Michalewicz, 1991; Rumble and Smith, 1991). The classical astronomical database consists of a table of highly structured records with fixed-length fields for numbers or text. These astronomical databases have traditionally been used to store observational data in original and processed form and they are in many respects very similar to databases found outside the sciences. Notable differences are missing values, the fact that the values in one table column may represent errors of the values in another column and, astronomy-specific, the various sky-coordinate systems.

Recently non-numeric and mixed information has been included into astronomical databases, particularly in the context of space-borne missions (Christian et al., 1992), such as instrument performance tables and documents, software documentation, visitor guides, observing schedules, and calibration target lists. These astronomical databases tend to consist of collections of semi-structured or unstructured files.

4.2.1 Requirements

As noted by Davenhall (1991) and Page (1992) the operations carried out on highly-structured astronomical databases are remarkably similar and include inserting, deleting, searching for, sorting and grouping rows, calculating columns, joining tables, and reporting results.

Page (1992) has compiled a rather comprehensive list of desiderata for a generic astronomical database for structured data. The list includes the quest for generic data import/export facilities, for exploratory data analysis capabilities (e.g. browsing and iterative selection of subsets), for the treatment of missing values, and a callable programmer's interface. According to Davenhall (1992), none of the existing commercial DBMS packages comes close to fulfilling all the desiderata. For a notably different view, however, see Benvenuti (1992).

4.2.2 Internal structure

While the internal structure of the classical astronomical database is that of a flat table, some are fairly complex. The prime example in the latter category is the HST science data catalog, stored in a relational database with 11 main relations and 6 so-called "master tables" which contain detailed engineering and instrumental parameters (Schreier et al., 1991).

Many astronomical databases internally rely on a commercial database management system, usually of the relational type. Others such as SIMBAD, EXOSAT or the Isaac Newton Group/La Palma/Westerbork archive use homegrown DBMSs, again mostly relational ones.

Several astronomical databases are offered as DBMS tools. The ESO-MIDAS image processing system, for instance, has for quite some time included a (flat) table file system tool (Ponz and Murtagh, 1987; Péron et al., 1992). The MIDAS table file system has served as the basis for a similar package implemented in the STSDAS image processing system (Hanisch, 1989, 1991). A versatile database system (Landsman, 1992) has also been integrated into the general purpose image processing system IDL (Research Systems, 1991). Astronet's DIRA database software system (Benacchio, 1991; Benacchio and Nanni, 1992) offers capabilities for centralized "system" databases and for personal databases; the latter are conceptually identical to system databases, except that they are managed by individual users.

A few DBMSs are intimately related (and tailored) to the astronomical information they manage. In this category we find SIMBAD (Egret et al., 1991), the EXOSAT database (Giommi and Tagliaferri, 1990; White and Giommi, 1991), and the ROSAT database (Zimmermann and Harris, 1991).

4.2.3 Design goals and design methodology

The goals set up in the design of the different astronomical databases vary from implementor to implementor[1]. In the design of SCAR, a relational DBMS in the Starlink software collection, *flexibility* was considered most important, followed by speed and efficiency in accessing the data (Davenhall, 1991). The designers of STARCAT (Russo et al., 1986; Richmond et al., 1987; Pirenne et al., 1992), the current UIF to the Hubble Space Telescope data archive and also to the ESO Archive (Ochsenbein, 1991; Albrecht and Grosbøl, 1992), have repeatedly stressed *simplicity* as a guiding principle. For the EXOSAT DBMS emphasis was placed on *graphics* and multi-table correlation capabilities (White and Giommi, 1991).

In the development of the EXOSAT DBMS, a user-centered design methodology was used by seeking active participation of the astronomical community (White and Giommi, 1991). The implementors of the NASA/IPAC Extragalactic Database (NED) invite user feedback even continuously, since "to underline the importance of the users' suggestions and opinions, every screen offers the user the option of leaving comments on the system" (Helou et al., 1991).

4.2.4 Query modes and languages

It is well recognized (Helou et al., 1991) that "the user's view of a database is strongly colored by the ease with which one interacts with it. This ease is defined by several factors, including: the amount of learning needed before sessions become productive; the degree of overlap between the specific questions a user has and those the database can answer; the power and flexibility available to the user in formulating a query; and the convenience of the mechanics of submitting a search and collecting the results. While some of theses factors are determined by the internal structure of the database, they can be modified by, and in the end depend mostly on, the user interface to the database."

Pasian and Richmond (1991) offer a similar view in stating that "the user interface is a fundamental component of information systems" and that "the extent to which information systems are useful is critically measured by how easy they are to use for both novices and experts."

The principal means of interacting with a database remains a query language of some form. Astronomical databases increasingly rely, at least internally, on the Structured Query Language (SQL; Date, 1987). For some astronomical databases, however, discipline-specific query languages have been or are being developed, which almost invariably have taken the form of imperative ("command") languages. A notable exception to this rule is the UIT DBS (Landsman, 1992), which offers a functionally-oriented language – in fact with the same syntax as IDL itself. For discussions of how to choose an appropriate database query language see Jarke and Vassiliou (1985) and Ozsoyoglu and Ozsoyoglu (1985).

---

[1] Of course, every implementor aims at user-friendliness and claims to have achieved it.

STARCAT offers essentially two query modes[2], namely by "commands" and query-by-example through forms. Associated with each form is a pre-canned SQL-query for which STARCAT constructs the WHERE-clause from the user's input. The SQL-query automatically performs all required relational joins. (However, no query optimization is carried out to remove unnecessary joins.) Forms are also used to display retrieved results record by record. STARCAT saves user actions as commands in a log-file, which can later be re-executed. The STARCAT commands effectively form a restricted "hierarchical" programming language (Pirenne et al., 1992).

The interface to the NASA/IPAC Extragalactic Database (NED), also principally based on forms, incorporates a sophisticated and versatile interpreter for object names, which internally uses the "lex" regular expression parser (Helou et al., 1991). NED users can query for objects by specifying an object name or a neighbourhood of some coordinates.

The joint Isaac Newton Group/La Palma/Westerbork archives use a discipline-specific ARCQUERY (archive query) language which is equipped with context sensitive help (Raimond, 1991). ARCQUERY produces an output table (in the same format as the catalogue) which will be used as input table by the next command unless explicitly told not to do so. ARCQUERY also lets the user save his/her personal default settings for a subsequent session.

SIMBAD, for instance, can be queried via three different modes (Egret et al., 1991). The appearance of the retrieved data is controlled by user specified formats which can be stored and reused in a later session.

In the Starlink Catalogue Access and Reporting (SCAR) system objects may be selected according to different sorts of criteria: simple selections, objects that fall inside an arbitrarily shaped polygon, or an arbitrary number of objects with the largest or smallest values for some specific field (Davenhall, 1991).

4.2.5 On-line help, tutorials, etc.

The amount and organization of documentation astronomical databases offer about the database contents and about the system's functionality varies substantially from system to system.

The designers of NED, for instance, stress that an essential feature of its interface is self-documentation with on-line help, longer explanations of the functions and their usage, an introduction/overview, a general tutorial, recent news and a glossary (Helou et al., 1991).

In my experience, astronomical databases tend to offer rather little help for accomplishing the fundamental task of mapping (usually imprecise) questions onto (usually formal) queries the system can handle.[3]

[2]Additionally STARCAT offers the option of viewing and modifying the SQL query it generates; however essential information about database structure and content is hidden from the user.

[3]It might be interesting to pursue the idea of compiling a set of natural language sample questions together with corresponding formal sample queries. These examples, after being indexed using WAIS, could be queried using natural language, and a WAIS proximity search would find one or more appropriate queries matching closely the given question.

4.2.6 Heterogeneous/distributed DBMSs

The many (often unnecessary) differences in the UIFs to the astronomical databases presented above lead to many (likewise often unnecessary) difficulties for an astronomer trying to access more than one astronomical database. In order to overcome these difficulties, efforts are underway both in the United States and in Europe to integrate the individual astronomical databases under single "umbrellas".

ESA's European Space Information System (ESIS) project aims at building a front-end for a heterogeneous, distributed database system (Albrecht et al., 1987; Albrecht et al., 1988). ESIS is envisaged to offer general support tools for information manipulation. These functions shall include a "scientific spreadsheet" and means to support user notes, glossaries and lexicons. Special tools are envisaged to support coordinate system and physical unit transformations (Albrecht, 1991).

NASA's Astrophysics Data System (ADS: Pomphrey and Good, 1990; Murray et al., 1992) program similarly aims at providing astronomers with "easy access to astrophysical data and analysis capabilities without regard to the location or sub-discipline orientation of the user, the data, or the processing tools involved" (Weiss and Good, 1991).

The main promise of these activities is transparent access to a set of heterogeneous, distributed astronomical archives through a standardized UIF with associated reduced learning efforts on the part of its users and also reduced software maintenance overhead at the users' sites.

4.3 UIFs for Unstructured and Semi-Structured Databases

Access to bibliographical references is vital for researchers in any scientific discipline. When references are organized in a digital database they become an invaluable and convenient source of information.

Compared to other scientific disciplines, notably mathematics, biology and computer science, astronomy is lagging behind in providing computer-aided, state-of-the-art access to bibliographic information. The formidable task of abstracting practically all astronomical publications is being carried out by the Astronomisches Recheninstitut Heidelberg, and results in several annual volumes of Astronomy and Astrophysics Abstracts (A&AA). Despite the fact that A&AA has been prepared electronically for several years (cf. Adorf and Busch, 1988), the abstracts are still not being made accessible on-line (or on computer diskettes) in a convenient form. The hopes of ever seeing A&AA, this single most precious source of astronomical bibliographical information, appear on-line as a separate and complete database are fading now (Watson, 1991).

Thus for computer-assisted bibliographical information retrieval astronomers have to turn to other, unfortunately less complete information sources such as INSPEC, ISI's Current Contents, or the NASA/RECON service (see below).

4.3.1 Local bibliographical database systems

ISI Current Contents on Diskettes

The Institute of Scientific Information (ISI) is a commercial enterprise engaged in scientific abstracting services. It produces Current Contents, a weekly database of titles, author names and addresses, keywords, and optionally abstracts of scientific publications. Originally being paper-based, the database has recently become available on computer diskettes, both in PC and Macintosh format. Abstracted journals are grouped into subscription sets, of which the one most relevant to astronomy is PCES (physics, chemistry, and Earth sciences[4]) which includes the major astronomical journals.

Subscribers to the service receive two or three new diskettes on a regular (weekly) basis. The records in each database issue can be accessed in two major modes, BROWSE or SEARCH. In BROWSE mode, publication titles can be viewed in the order in which they are stored on disk, or by discipline or by journal name. Each publication can be viewed either individually or in a list of authors and titles. Publications of interest can be marked for inclusion in a "personal interest" list or in a "request a preprint" list. Whenever an article is selected, visual clues indicate whether and where it has been included. Detailed information on an article's author or the publisher can be displayed by a simple button click. Other buttons allow e.g. to print all records of interest.

In SEARCH mode (Figure 4.1) a form is used to specify a query called "search profile". Each line in a search profile permits to specify allowed values for a single field, which may be either the journal's name, or the author of the publication, or its type etc. or just everything. The query (or the set of records retrieved, depending on one's perspective) receives a unique "set number" which can be referred to in further queries involving a special "set number" field. Permitted values can be combined by the ordinary Boolean logical connectives AND, OR, NOT and parentheses.

The formulation of a query is aided by several data dictionaries, i.e. lists of index terms which can be field values. There is one dictionary for each field and dictionaries are specific to each issue of Current Contents on Diskette. Each dictionary can be accessed via a pop-up menu and along with the field values displays the number of matching records. Single or multiple field values can be selected from the pop-up menu for inclusion in the query statement. The query line can be edited in the usual way using the mouse pointer, the forward and backward delete keys, the arrow keys and cut-and-paste operations.

Upon completion of the query specification, the system immediately consults the database and, while the user contemplates the next query line, returns the number of matching records found. Thus the user can formulate a set of partial queries and the system provides useful feedback on the population of the database. From the partial queries one or more final queries can be composed using set numbers together with Boolean connectives.

Search profiles can be saved into disk files and re-used, e.g. on a different database issue. Several profiles can be run one after another without mutual interference, since set

[4] A free evaluation copy of the required retrieval software and some sampler disks can be obtained from ISI.

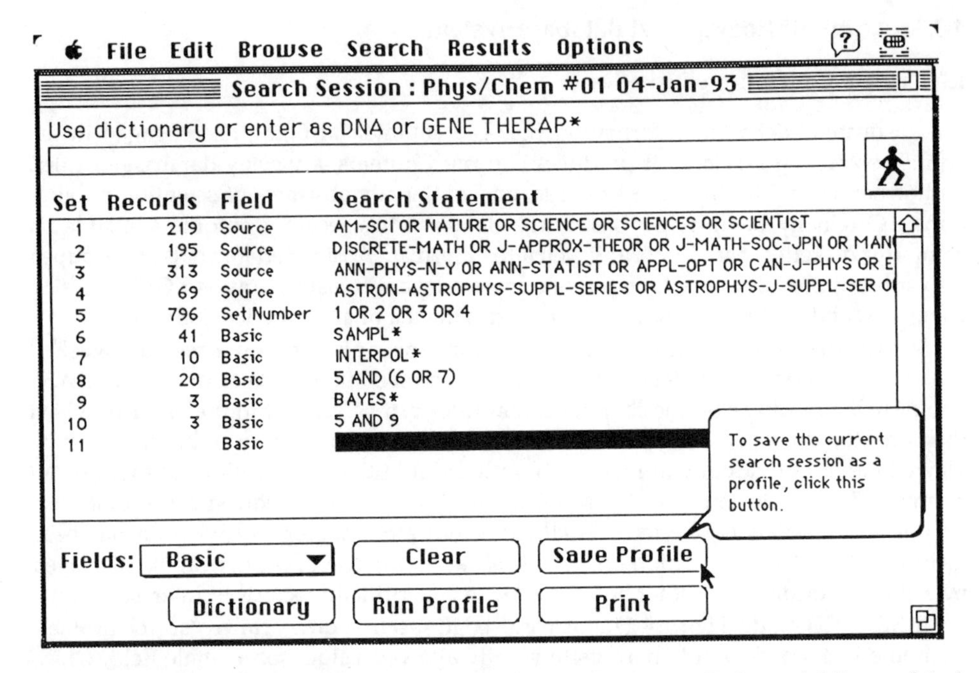

Figure 4.1: The user-interface to the Current Contents on Diskette bibliographical database on the Apple Macintosh. Sets of records selected by the search statement (query) are automatically numbered. The number of records in the set is displayed along with the name of the field to which the search statement applies. Different sets can be combined using Boolean connectives. A double mouse click on a set displays the set as a list of headlines or record by record.

numbers are relative, not absolute.[5]

A valuable feature of the ISI Current Contents is its cross-disciplinarity. Thus the same search profile can be run against a set of journals from mathematics, statistics, physics, and astronomy simultaneously. A potential problem, however, is that different terminology may be used in the various scientific disciplines.

The ISI Current Contents DBMS cannot be used to delete unwanted records from the database or to import foreign records. However, the DBMS permits to export selected subsets of records to a disk file in one of several formats including simple ASCII or tab-delimited text, Pro-Cite, EndNote and Dialog. Thus bibliographic information can be readily imported into a personal bibliographical DBMS such as EndNote (see below) with just a few mouse clicks.

[5]Unfortunately, the sets cannot be given user-defined symbolic names, so a query cannot relate to the results of e.g. two previous queries.

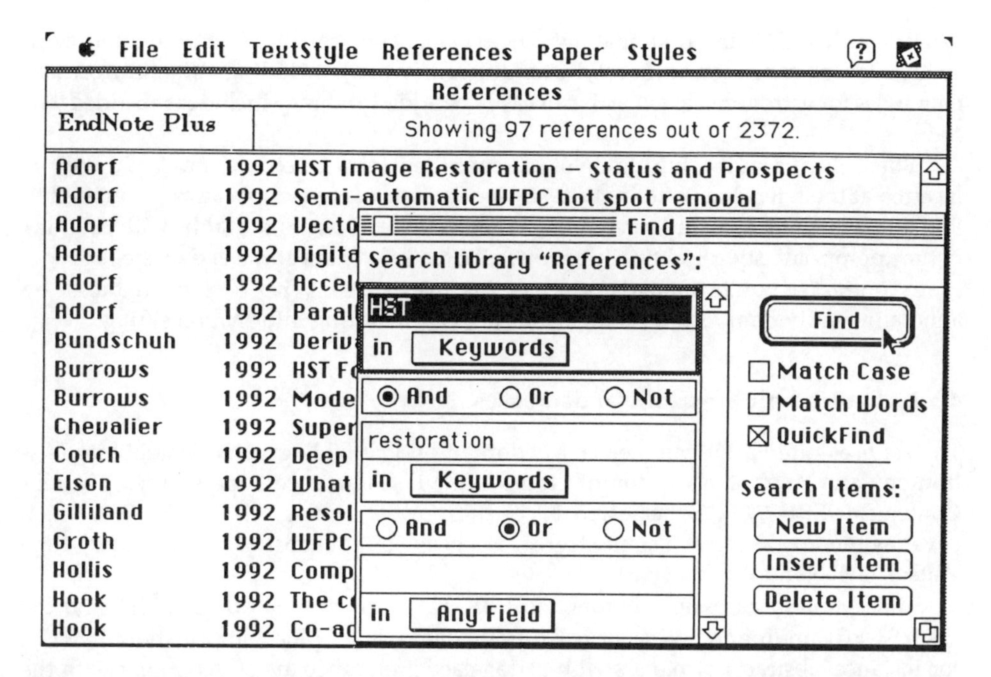

Figure 4.2: The user-interface to the EndNote bibliographical database on the Apple Macintosh. References can be selected by specifying restrictions to the values of the database fields. The selection is displayed in list form. Records are mouse-sensitive and can be opened or exported with a few mouse clicks.

EndNote

EndNote from Niles and Associates is a commercial personal DBMS tool explicitly customized for storing and retrieving scientific bibliographical records. While in principle practically any database system can be used for bibliographical purposes, specialized bibliographical DBMS software such as EndNote incorporates knowledge about bibliographic subtleties such as abbreviation and positioning of author initials, data dictionaries for journal names, different styles and abbreviation levels required for different journals, etc.

In EndNote, bibliographical records, which are of variable length, can be selected using a menu and Boolean connectives (Figure 4.2), though, in the absence of parentheses, not all possible Boolean queries can be formulated. Some fields such as author and keywords are pre-indexed for faster retrieval. The user has some control over the matching process (case sensitivity, exact vs. partial match). The selected records can be sorted in ascending or descending order using multiple keys mapped to fields.

EndNote, particularly in conjunction with its extension EndLink, can import records from a variety of file formats including those produced by on-line bibliographical database services such as STN International/INSPEC, Dialog, and Pro-Cite. EndNote comes with about a dozen predefined output styles for different journals including *Nature* and *Science*; new styles for astronomical journals can easily be added and may include embedded TeX code.

When, during authoring, EndNote is consulted for a reference, an ASCII stub is inserted at the reference point in the text. The finished compuscript can be submitted to EndNote which looks up all references in the database, replaces the stubs with citations in the appropriate style and finally collects all cited references at the end of the text.

In contrast to public databases, the user can customize his/her personal database by annotating individual records, a highly useful feature during information sifting.

4.3.2 On-line bibliographical databases

The services offered by commercial vendors of on-line bibliographical databases (Astronomy and Astrophysics Monthly Index, CONF, Conference Papers Index, Current Contents, INSPEC, PHYS, SciSearch, NTIS, BRS, ORBIT, Dialog, STN, ESA/IRS), which cover astronomy at least to some degree, has been comprehensively reviewed by Rey-Watson (1987) and Watson (1991).

Some publicly accessible astronomical databases, such as SIMBAD, NED, and the ESO/ST-ECF on-line archive, also include valuable bibliographical information. In NED, for instance, abstracts of papers with extragalactic relevance are stored along with the objects they refer to. This effort has resulted in 1,051 and 1,140 abstracts for 1988 and 1989, respectively, requiring roughly 2.5 Mbyte of storage per year (Helou et al., 1991). Corbin (1988) describes a project of archiving past observatory publications in a full-text database on optical disks.

However, the earliest and most comprehensive of databases encompassing the entire field of space and space-related science including astronomy, is the NASA/RECON database, which dates back to 1962 (Watson, 1991). Recently the RECON database has been made publicly available as part of the STELAR project (Van Steenberg et al., 1992) and, since it can be conveniently accessed via WAIS (see below), it can be considered today's most valuable on-line source of general astronomical bibliographical information.

4.3.3 The wide-area network servers

Information services on wide-area networks have recently sprung into existence and are experiencing an almost explosive growth in number and usage. Among the most popular services one finds the Internet Gopher, the World Wide Web (WWW) and the Wide-Area Information Servers (WAIS) system[6] (Chevy, 1992; Krol, 1992; Press, 1992a, b). WAIS operates according to the client/server principle and provides a homogeneous user interface (the client) to a large number of document databases (the servers). WAIS

[6]WAIS was invented to promote massively parallel machines (particularly TMC's Connection Machines) which are capable of performing a free-text search of an 8 Terabyte archive in about 15 seconds using a pre-computed inverted index.

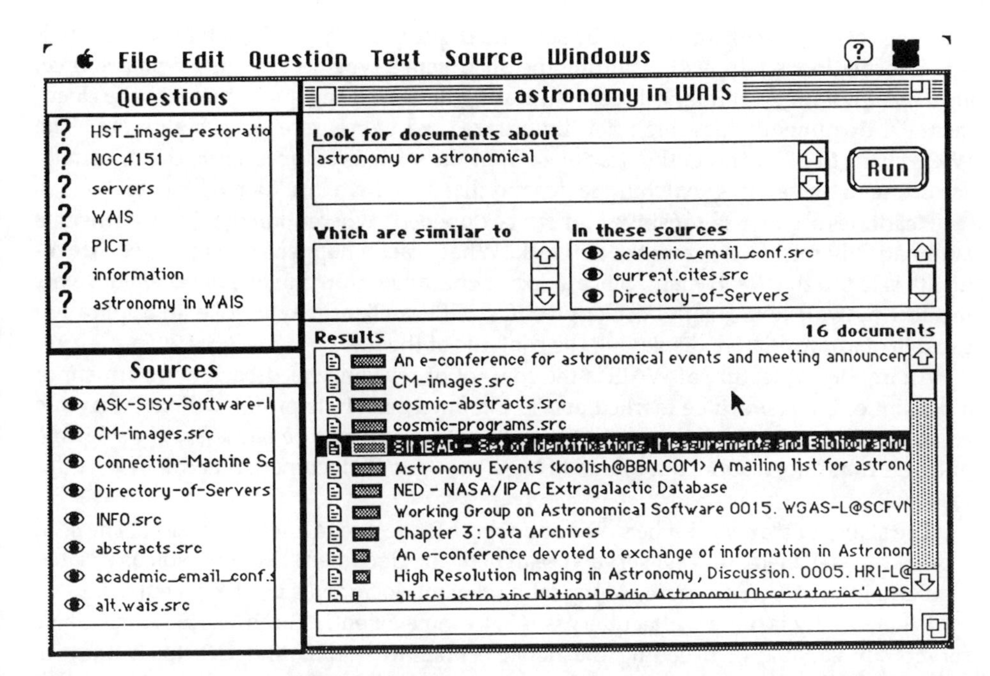

Figure 4.3: The Apple Macintosh WAIStation user-interface to the wide-area information services (WAIS). Saved WAIS questions are to be found in the upper left corner window, known WAIS sources (servers) in the lower left corner window. The "astronomy in WAIS" question is opened and, after a search in the specified sources (upper right panel), a list of document headlines is displayed (lower panel) along with an icon for the document type and the matching score. A double click on the headline retrieves the associated document from the remote database server.

client, server and indexing software for a variety of hardware platforms is available free of charge via anonymous ftp from `think.com`.

WAIS is truly cross-disciplinary. When searching for information using WAIS one often retrieves headlines of documents one has not directly asked for. This should not be viewed as a pure nuisance, since these documents might contain interesting related information either from one's own discipline or a neighbouring one.

In order to carry out a WAIS search, the user interacts with a local, usually graphical WAIS client (Figure 4.3). A few seed words or a whole natural language phrase are entered into a "question area". Next, one or more information "sources"[7] to be used in the search are added in a "source area" and then the query is submitted.

The WAIS client connects to one or more remote WAIS servers which retrieve the

[7]A source description contains network access information for a particular WAIS server and a brief documentation of the contents of the database in question. Source descriptions must be available locally; they can be downloaded from a centralized "directory-of-servers" source.

headlines of documents matching the submitted query. Along with each headline auxiliary information is retrieved including the document's type, its size, its original source, and, most notably, a numerical score indicating how well the query matches the document. A document with a high density of rare seed words receives a high score. The WAIS client finally collects the headlines from the various sources, sorts them, usually according to descending matching score, and displays them in a "document area".

Headlines are mouse-sensitive and can be "opened" whereupon the full document is automatically retrieved across the network. What exactly happens to a retrieved "document" is left to the WAIS client. Since a document can be stored in one of several formats including ASCII text, rmail, PostScript, PICT, TIFF, a client may invoke an application capable of appropriately displaying the contents of the file.

An important feature of WAIS is the concept of relevance feed-back: a document or a portion of interest can be marked and in a refined search its profile will be taken into account. The purpose of the initial query is not so much to be successful in selecting only relevant information, but to find at least one "good" document that can be used for feedback.

The status of a query can be saved as a whole for re-use in another session. The seed words, the specified sources (servers) searched, the document headers plus associated auxiliary information are stored along with relevant documents used for feedback

What exactly is done with a query is left, to some extent, to each WAIS server. Some servers are capable of decoding the Boolean operators AND and OR, to distinguish between a collection of words and a proper phrase, and to restrict the search to controlled keywords or other sections of a document; other servers are less sophisticated. In any case, all documents stored by a server are pre-indexed in order to increase the efficiency of the retrieval process.

For astronomers the currently most interesting WAIS server is presumably NASA's RECON database (Warnock et al., 1992) already mentioned above, which contains abstracts for seven major astronomical journals (AJ, ApJ, ApJS, A&A, A&AS, MNRAS and PASP). The RECON abstracts differ from those provided by the authors of the articles.

4.4 UIFs for Highly Structured Databases

Highly structured databases are the domain of classical DBMSs. Below we will inspect a spreadsheet, two graphical exploratory database tools and a DBMS for large databases

4.4.1 Spreadsheets

Spreadsheets are widely used in the business world for data storage and analysis, but they are relatively little known in scientific circles. A spreadsheet is primarily a visual interface for rectangular data tables permitting easy direct manipulation. For instance, once stored in a spreadsheet, data columns and rows can easily be rearranged. Columns and rows can be hidden in one or more hierarchical levels. In addition to their interactive manipulation capabilities, spreadsheets can be programmed.

Spreadsheets are memory based. Their fundamental data structure is an array (or matrix) of "cells" named A1, A2, ..., B1, B2, ... at the cross-sections of columns A, B, ... and row 1, 2, ... Cells, and groups of cells, can also be given mnemonic names. Individual cells roughly correspond to variables of conventional programming languages. However, as opposed to static languages such as Fortran or C, spreadsheet variables are dynamically typed: a cell derives its data type from the value it holds. Valid data types comprise integers, floats, character strings, but also vectors. Cells are also dynamically formatted: different formats can be defined for different possible data types; the appropriate format is selected according to the current data type of the cell. Dynamic typing and formatting greatly facilitates interactive work.

The concept of a spreadsheet differs in another aspect from that governing conventional programming languages: a spreadsheet holds data and program (see Appendix 3) together in volatile memory. The memory state can be saved as a whole to disk and restored at a later time. Everything, including the cursor position and the scrolling to a particular area, is restored, so that the user can continue his/her work at a later time as if it had never been interrupted. Data persistency across multiple sessions is an active research topic in computer science (see e.g. Sajeev and Hurst, 1992).

In the following I shall describe some features of a particular, widely used spreadsheet, namely Microsoft Excel on the Apple Macintosh; the package is also available on personal computers such as IBM PCs. Other spreadsheets, notably Lotus 1-2-3, offer comparable functionality.

Graphical displays

The real power of spreadsheets for exploratory purposes stems from their integrated graphics. One or more data columns (or selections thereof) can be charted in scatter- or line-graphs. Charts can be created in separate windows or as an insert in the same window. Each chart is dynamically linked to the worksheet from which it is derived. If a value changes on which a chart depends, the chart is automatically updated.

Accessing remote databases

For quite some time MS Excel has been offering the option for defining an internal database, consisting of an area of rows (the records), in which records could be searched, modified, extracted and deleted. The search is controlled by one or more search criteria specified in spreadsheet rows. Criteria in the same row are ANDed and criteria in separate rows are ORed.

The most recent version of MS Excel (V4.0) has extended this functionality towards external databases. The user can choose from one or more remote hosts and one or more database systems available on these. After establishing a connection to a remote database, the user, with the help of a menu, can specify the field names for which information is requested (Figure 4.4). Field names are automatically pasted into the spreadsheet in the order in which they were specified. Next the query is formulated in one of several ways: search criteria can be specified directly in the spreadsheet. (Even relational joins across

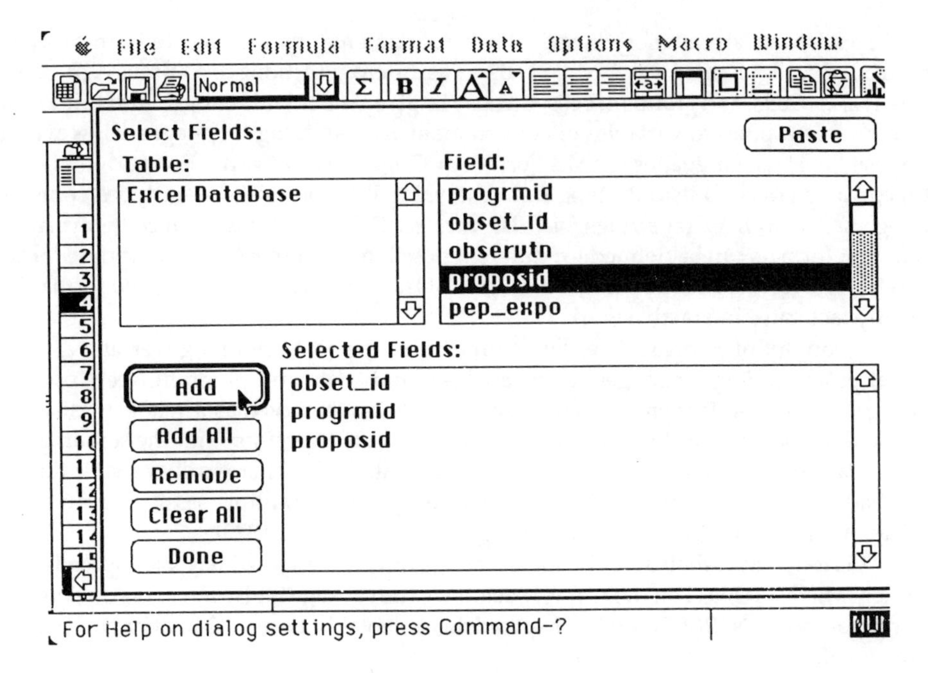

Figure 4.4: Query formulation in the user-interface to Microsoft Excel on the Apple Macintosh. The fields available in the selected database (upper left panel) are displayed in a menu (upper right panel). Fields can be added to the list of selected fields (lower panel) by a mouse click. They are later automatically pasted into the spreadsheet.

several tables can easily be formulated.) Alternatively, the user might seek guidance by a query assistant. In any case the result is an ordinary SQL-query, which may be inspected and modified. The query is submitted and the retrieved records are received directly in the spreadsheet for inspection or further analysis.

Technically speaking, MS Excel uses Apple's Data Access Language (DAL), a derivative of SQL, for communicating with a remote database. DAL operates according to the client/server concept. The spreadsheet passes the query to a Macintosh DAL-client, which in turn communicates via the network with a DAL-server on the remote host. The latter provides the necessary information to access the DBMS, to submit the query and to receive the result, which is transferred back to the DAL-client on the Macintosh and further passed on to the spreadsheet. The common relational DBMSs such as Oracle, Ingres and Sybase can be accessed this way.[8]

[8]The spreadsheet software is accompanied by two HyperCard help stacks which help to familiarize with the concepts and features involved. Within the spreadsheet itself hypertext-based help information is provided at any stage. Remarkably, given an error, the user is automatically "thrown" into an appropriate section of the help.

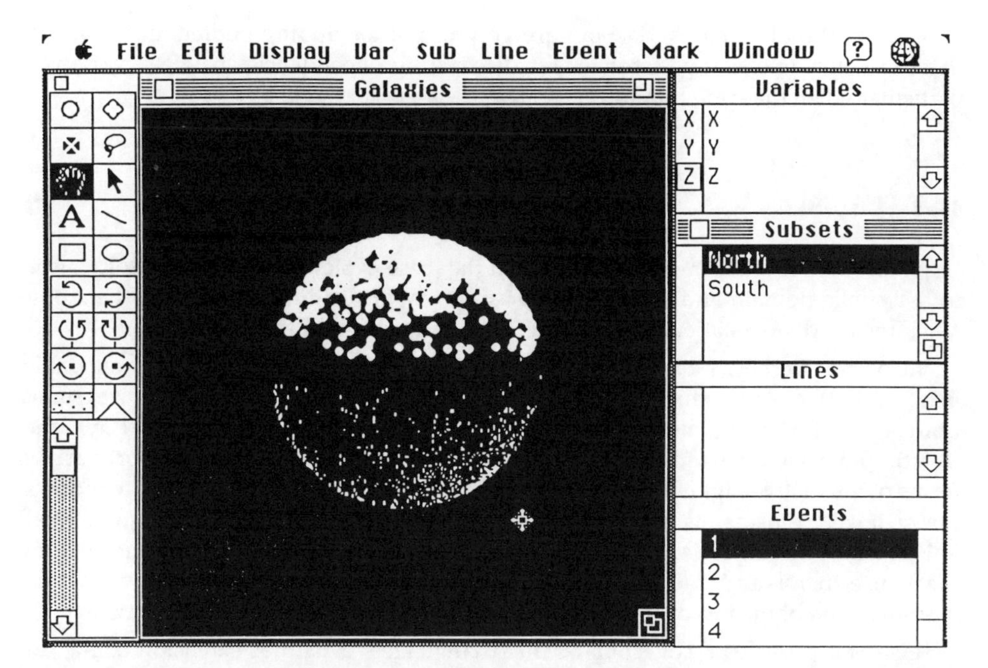

Figure 4.5: The user-interface to the MacSpin exploratory database analysis tool. The graphical display of a set of records (middle panel) can be rotated using a variety of tools (left panel) including a "hand". Variables (i.e. fields) can be selected for display (top right panel). Subsets can be selected and named (second panel from top). Events (i.e. records) can be individually selected (bottom right panel).

4.4.2 MacSpin

MacSpin from D2 Software Inc. is a commercial graphical data analysis software which can be used to advantage for exploring numerical multi-column data tables. Each record is represented as a point in a high-dimensional dataspace (Figure 4.5). The software allows the identification of a database record in a graphical display by pointing at it with the mouse. Also slices through a point cloud can be displayed.

Various mouse tools (including a "lasso") can be used for selecting records. Different selections can be marked with different symbols and/or colour. The user can focus on a particular selection and save it as a named subset. Subsets can be combined into new subsets using the set exclusion, union and intersection operations.

The point cloud can be rotated automatically or by a "hand" tool in order to view the data from different perspectives. For data sets of moderate size the rotation is fast enough to give the user the impression of a dynamic system, thereby exploiting the brain's capability to add a third dimension where in fact only two are displayed on the screen.

Graphical tools such as MacSpin are very useful for finding outliers in numerical databases, for detecting patterns, or for discovering dependencies between variables, particular when these are non-linear.

4.4.3 Lisp-Stat

Lisp-Stat can be considered as an advanced database exploration tool. The software is in the public domain and is available for a variety of platforms, including the Apple Macintosh and workstations running the X11 window system under BSD Unix.

Lisp-Stat (Tierney, 1988, 1991; Chambers, 1989; Lubinsky, 1991) is based on David Betz's XLISP (see Appendix 2), which in turn is implemented in C. Lisp-Stat's power stems from several forms of plots, including histograms, scatterplots, 3-dimensional dynamic (rotating) plots and also scatterplot matrices. These graphical displays support various forms of interactive highlighting operations and, when displayed in different windows, can be linked; thus points highlighted in one display window will be highlighted in all linked windows (Figure 4.6). Interactions with the display windows are controlled by the mouse, menus and dialog boxes. An object-oriented programming system allows the customization of menus, dialogs, and the way windows respond to mouse actions.

Lisp-Stat provides a collection-oriented language – a natural extension of the underlying Lisp-language – and is quite a useful tool for exploring the multi-dimensional dataspace spanned by the columns of a database table. The system is versatile, since the user can utilize a fully-fledged programming language.

4.4.4 The UIT Database System

The UIT Database System is a public-domain database package integrated into the commercial IDL image processing system. The software, designed by Don Lindler, is noteworthy for its speed and versatility which arises since the database software shares IDL's programming and plotting capabilities (Landsman, 1992). The UIT DBS package offers a total of 35 DBMS procedures, of which six core procedures will often be all that a user requires. The UIT DBS can be used for both system-wide database and personal databases. Every field can be indexed, even numerical ones!

The design of the UIT DBS is collection-oriented in the following sense: the result of a query (search) operation is not a list of records, but a set of pointers to the selected records. This set can be fed back in a subsequent query, thereby speeding up the query refinement process.

A pointer set can be used to control the extraction of some table columns which are simply bound to ordinary IDL variables for subsequent processing and/or graphical display. Conversely a database may be populated by storing the values of ordinary IDL vectors into the columns.

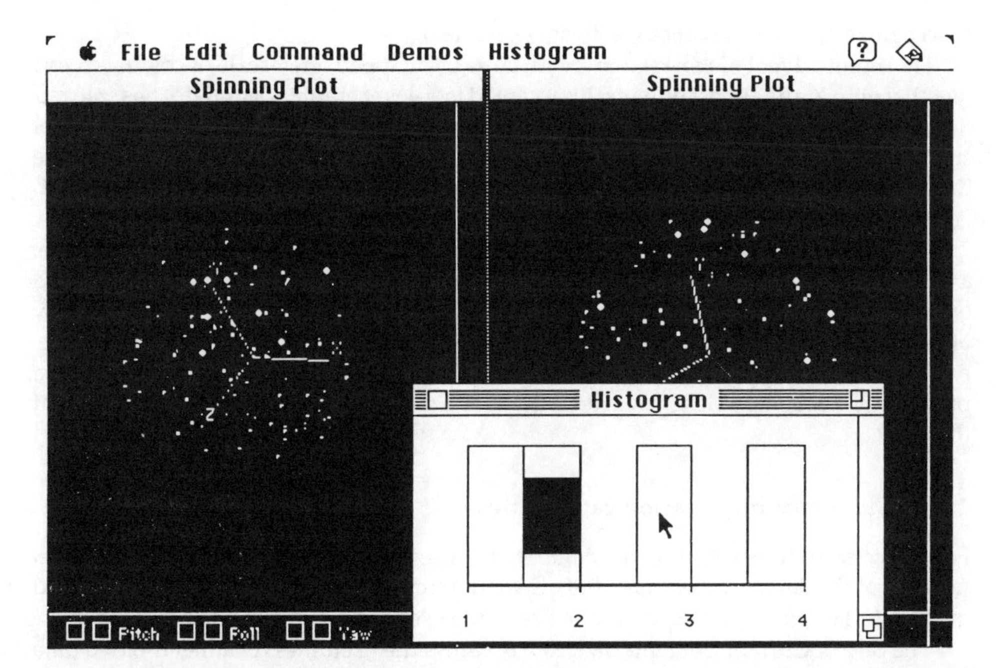

Figure 4.6: The user interface to the Lisp-Stat tool for database exploration. Two spinning-plot windows are used for data display. A third histogram window in the front is used to select records (points in the spinning-pot windows) from the dataset. All windows are linked. Thus a selection done in any of them is immediately visible in all others.

4.5 Useful Concepts

So far we have inspected a number of commercial and public-domain "front-ends to databases" with the aim of identifying concepts which facilitate information retrieval and manipulation activities. Some of these features enhance the user-friendliness of the application, others provide sophisticated functionality. Below, a few of the concepts are further discussed. It is envisaged that more and more of these features will be incorporated into astronomical databases in the course of time.

4.5.1 A direct-manipulation graphical user interface

There is consensus that in most cases a graphical user-interface (GUI) greatly facilitates the human interaction with a database, particularly for the novice or the occasional user. The idea of GUIs, originally developed in the Xerox PARC artificial intelligence research laboratories, was widely popularized by the Apple Macintosh (Marcus and van Dam, 1991). Graphical user interfaces have a much larger information transmission capacity than ASCII-based screens, since they better match the human cognitive capabilities tuned to recognize real-world scenes (i.e. images), not text.

The recent advent of workstations with large bit-mapped, grey-scale or colour screens together with X, the almost universally accepted low-level graphics standard, has spurred the development of GUIs for science computing in general. However, their look-and-feel still varies too much across applications and thus workstation GUIs often do not generate the "homey" feeling as most applications on the Apple Macintosh do right from the start. For a checklist on good UIF design see Brackeen et al. (1989).

The example of MS Excel shows the usefulness of a direct manipulation interface, and of a facility for annotating individual fields, for highlighting certain rows, columns or single fields. It helps when the information density or, conversely, the readability of information presented on the screen, can be enhanced by controlling its appearance, e.g. font, size and typeface. Finally, MS Excel demonstrates the convenience of data persistency, i.e. being able to save the state-of-affairs to a disk file for later resumption of the work.

4.5.2 Summary information capabilities

For a successful interaction with a database the user has to form a coarse mental model not only of the database front-end, but also of the contents of the database. A form-based query interface, for instance, is of little use, if the user has no idea of what to fill in.

The case of Current Contents on Diskette shows the usefulness of a menu-based data dictionary providing a quick overview of which values may occur in a particular text field. If the number of possible values is too large to be displayed in a simple sorted list, a hierarchical organization (menu with sub-menus) is advantageous.

For a column containing numerical values a data dictionary simply listing all values would be of little use. The contents of such a column have to be represented differently, e.g. by a histogram or by a list of quantiles together, perhaps, with a list of outliers.

The interaction with a database is greatly facilitated if user input is "intelligently" parsed. A DWIM ("do what I mean") interface is particularly important for the "difficult" fields containing dates, times or sky-coordinates with their particular formatting conventions. Again MS Excel can serve as an example for a system that allows dates to be entered in almost any sensible format. The output format of these difficult fields should, of course, be controlled by the user, and not by the system's internal storage convention.

Multiple, linked dynamic views, as provided in Lisp-Stat, facilitate the process of forming a mental model of a database's contents, particularly for numerical fields. Colours and different marker symbols, as provided in MacSpin, facilitate the recognition of sets of interest in a multi-dimensional graphical representation. An attractive way of selecting record-sets obviously is the use of a mouse cursor for encircling a set of points each representing a database record.

4.5.3 A programmer's interface

Graphical user-interfaces are currently very popular, but they are not a panacea solving all human-computer interface problems. Many of them lack programmability, and therefore extensibility and flexibility.

Some GUIs, such as MS Excel, provide a so-called macro recorder which permits the application to be programmed simply "by doing". The macros are recorded and can be edited afterwards as any ordinary program. The STARCAT log-file capability provides a similar means to record actions as commands for later re-execution.

However, special purpose ad-hoc programming languages show problems, which can often be traced back to the history of these languages. (They were not designed as programming languages from the outset.) Each of them has a different syntax, requiring special parsers and documentation efforts on behalf of the system providers, and learning efforts on behalf of the users. Their special syntax also inhibits the porting of small routines, such as date-conversion utilities, across database applications. Most of these mini-languages are incomplete, and, being imperative, are not very expressive; they lack a foreign function interface to call routines written in other programming languages, etc.

Many of these problems would fade away, if the providers of astronomical databases could agree on a simple-to-read and simple-to-parse syntax for their "scripting" language and/or communication protocols (see Appendix 1). Here a list-oriented syntax, as adopted for the programmer's interface to the Emacs editor (Stallman,1987), for the extension language to the AutoCAD program for computer aided design (Hafner, 1992), and also for the WAIS source/query specification, offers many advantages, including language sensitive editing (e.g. in the Emacs editor). Moreover, due to its simplicity, list-oriented syntax is particularly easy to write programmatically.[9]

4.5.4 Aids for query formulation and query optimization

Some of the database front-ends described above offer more than one way of formulating a database query, thereby responding to the different needs and capabilities of different users.

MS Excel, for instance, offers three query mechanisms: a query-by-example facility using a form for simple queries; a query specification in the spreadsheet for more complex ones; and access to the full SQL-query string. In addition there is a "query assistant" module. MS Excel is noteworthy for its built-in facility for mapping the result of a complex SQL-query reflecting the structure of a relational database into a simple flat table. It achieves this, not by hiding information, but by displaying information about the database in an appropriate context.

Formulating a query for a complex relational database consisting of many tables is a difficult activity requiring extensive knowledge. Semmel (1991) has used artificial intelligence techniques (cf. Rosenthal, 1987; Adorf et al., 1988; Adorf, 1991) to construct a system for heuristically based automated query formulation. The system incorporates syntactic knowledge to ensure that the query conforms to the rules of the formal query

[9]Adopting a list-oriented syntax for data structures does not automatically mean adopting Lisp as the programming language, as is demonstrated in the case of WAIS where client/server programs are usually written in C. Even code that is specified in Lisp-style must not necessarily be executed by a Lisp system. Weerawarana and Wang (1992), for instance, describe a system that allows the specification of supercomputing algorithms in Lisp style, which is eventually translated to and executed in Fortran-77.

language used for retrieving information from the database. Semantic knowledge ensures that a syntactically correct query will generate the desired result from the underlying database.

Semmel (1991) discusses how his software can be used to create "intelligent" database interfaces (cf. Chang et al., 1988; Parsaye et al., 1990; Chignell and Parsaye, 1991) that generate query language code from high-level requests, thus minimizing the need for syntactic and semantic knowledge, and considerably simplifying the formulation process.

Semmel's Lisp code, or a translation thereof, will be used within the StarView interface (Johnston, 1992) developed at the Space Telescope Science Institute for the HST Data Archive and Distribution Service (DADS) archive. In a test of the prototype, an entity relationship model of the Space Telescope DADS was entered which consists of 27 entity types, 31 relationship types, and 260 attributes. In a test, the system successfully expanded a high-level request consisting of 8 lines into an SQL query consisting of > 100 lines.

4.5.5 Collection-orientation

As opposed to users of commercial database applications, users of astronomical databases often retrieve one or more sets of records. It is therefore important to be able to view such a set in a convenient form. EndNote and Current Contents on Diskette display a set as a list and a double mouse-click on any record will open a window displaying the full record.

Working with sets is greatly facilitated when multiple focus sets can be defined, as e.g. in the case of INSPEC or Current Contents on Diskette. The user further benefits from the ability to form subsets of sets and to combine sets using the usual set-operators, as in the case of INSPEC or Current Contents. For databases with many and long records it is advantageous if a set just consists of the pointers to the original records rather than being composed of record duplicates. Working with sets is facilitated by a collection-oriented programming language (Sipelstein and Blelloch, 1990).

When sets of numerical records are displayed as multivariate point clouds it is convenient to have mouse tools for discriminating objects and thereby selecting (or excluding) a set of records. Such toolsets exist, as we have seen, in Lisp-Stat and MacSpin.

4.5.6 Two-way feedback

Feedback in both directions is one of the most important characteristics of a modern database front-end.

During a query the user needs feedback information on the status of the search. Before a set of records is displayed, the user needs the information of how many records the set contains, as is the case with INSPEC and Current Contents. When a database with variable-length records is queried, it is valuable for the user to see the size of a record, before actually attempting to retrieve it.

Relevance feedback, exemplified in WAIS, is an important and amazingly powerful concept for textual information retrieval.

4.5.7 Tutorials, on-line help and graceful error recovery

A user-friendly database interface offers information about itself. Several commercial database applications are accompanied by tutorials presenting an overview of the capabilities the system offers. A case in point is the WAIStation client for the Macintosh which features a movie with soundtrack. Another example is the MS Excel spreadsheet offering two excellent hypertext tutorials, which allow the user to freely move around in the material presented, and to learn selectively with user-controlled presentation speed.

Hypertext and hypermedia systems (Adorf, 1989; Nielsen, 1990; Marcus and van Dam, 1991; Berners-Lee, 1992, Bogaschewsky, 1992) widely popularized by HyperCard on the Macintosh, represent the state-of-the-art for user-friendly help systems. Again, MS Excel can serve as a good example: relevant help sections can be searched via keywords; all terms on which information is provided are highlighted in the text and are mouse-sensitive. The user can additionally navigate by tracing his/her trail forwards or backwards.

Advanced help systems are event driven and can therefore be invoked at any instance. Properly engineered, they do not destroy the context from which they are called and present the help information in a separate window. A work-around for a system which does not retain the context is to start up two separate processes, each in its own window, and to use one for the database interaction and one for obtaining help.

4.5.8 Integratability

With an increasing number of databases offering information it becomes increasingly important to have some flexible means for querying these databases and for processing the retrieved information. The current tendency is to build closed database systems and super-systems such as ADS and ESIS which offer a number of predefined services. However, since it is difficult to exactly foresee the users' needs and to build a system general enough to cover all of these, an alternative, promising approach is that of providing an open "Lego box" of simple building blocks which can easily be assembled to specific modules of higher complexity (cf. Adorf, 1993).

The deficiencies of closed systems become apparent when the task consists in merging output from several stand-alone database applications, and to control later processing steps by results obtained in earlier ones. What is needed is either an extensible system[10] or a set of tools, ideally integrated into an environment, that allow further processing of the output from various sources.

A standard query language such as SQL cannot be used for these purposes. Being specifically designed to permit the formulation of interactive queries to relational databases, this language has serious deficiencies: firstly, it is not integrated with (and not seamlessly integratable into) the major scientific programming languages Fortran and C ("impedance mismatch"); secondly, it is not functional, i.e. an SQL-query does not produce a result which can easily be further processed programmatically.

[10] An example for an extensible sytem is the well known Emacs editor (Stallman, 1987), which can be customized via its programmable Lisp interface.

Clearly, the "glue" required in information retrieval, as in many other areas, can be provided by an interactive, expressive, extensible, standardized, high-level programming language. Among the high-level programming languages available today, Lisp (see Appendix 3) is certainly a viable choice for providing the glue between standalone applications and for acting as a common top-level.

In order to check the viability of this idea, a few experiments were carried out with a WAIS- and a STARCAT-server written in Common Lisp (see Appendix 4). These experiments are encouraging, but further work is certainly required in this direction.

4.6 Summary

A number of contemporary astronomical database systems have been reviewed from a user's point of view. For comparison several commercial and public domain front-ends to databases have been presented in order to detect design features that can facilitate information retrieval and filtering tasks. Several useful concepts were identified: a direct-manipulation user interface, a programmer's interface, aids for query formulation including summary information capabilities, tools for query optimization, collection-orientation, two-way feedback, hypertext-based on-line help and tutorials, and integratability.

It hoped that this review can serve to improve existing astronomical database management systems and to provide some guidance for the design of new ones.

Acknowledgements

I appreciate discussions with Miguel Albrecht (ESO), Mark Johnston (STScI), and Benoît Pirenne (ST-ECF), and the persistent encouragement of the editors, André Heck (Observatoire Astronomique de Strasbourg) and Fionn Murtagh (ST-ECF).

References

1. Adorf, H.-M., "Hypertext and Hypermedia Systems", *Space Information Systems Newsletter*, **1**, 7–14, 1989.

2. Adorf, H.-M., "Artificial Intelligence for Astronomy", *The ESO Messenger*, **63**, 69–72, 1991.

3. Adorf, H.-M., "Reflections on the second Astronomical Data Analysis Software and Systems conference", *ST-ECF Newsl.*, **19**, 22–23, 1993.

4. Adorf, H.-M., Albrecht, R., Johnston, M.D. and Rampazzo, R., Towards Heterogeneous Distributed Very Large Data Bases, in *Astronomy from Large Databases: Scientific Objectives and Methodological Approaches*, F. Murtagh and A. Heck, eds., European Southern Observatory, Garching, 137–142, 1988.

5. Adorf, H.-M. and Busch, E.K., "Intelligent access to a bibliographical full text data base", in *Astronomy from Large Databases: Scientific Objectives and Methodological Approaches*, F. Murtagh and A. Heck, eds., European Southern Observatory, Garching, 143–148, 1988.

6. Albrecht, M.A., "ESIS – a science information system", in *Databases and On-Line Data in Astronomy*, M.A. Albrecht and D. Egret, eds., Kluwer Academic Publishers, Dordrecht, The Netherlands, 127—138, 1991.

7. Albrecht, M., Russo, G., Richmond, A. and Hapgood, M., "European Space Information System pilot project, user requirements document", ESRIN, Frascati, Rome, 1987.

8. Albrecht, M.A., Bodini, A. and Pascual, J., "Towards a European Space Information System – an evolutionary approach", *ESA Bulletin*, No. 55, 64–67, 1988.

9. Albrecht, M.A. and Egret, D., *Databases and On-Line Data in Astronomy*, Kluwer Academic Publishers, Dordrecht, 1991.

10. Albrecht, M.A. and Egret, D., "From archives to information systems in astronomy", in *Astronomy from Large Databases II*, A. Heck and F. Murtagh, eds., European Southern Observatory, Garching, 17–27, 1992.

11. Albrecht, M.A. and Grosbøl, P., "The ESO archiving facility: the 1st year of archive operations", in *Astronomy from Large Databases II*, A. Heck and F. Murtagh, eds., European Southern Observatory, Garching, 169–172, 1992.

12. Belkin, N.J. and Croft, W.B., "Information filtering and information retrieval: two sides of the same coin?", *Communications of the ACM*, **35**, 29–38, 1992.

13. Benacchio, L., "Database applications in Astronet", in *Databases and On- Line Data in Astronomy*, M.A. Albrecht and D. Egret, eds., Kluwer Academic Publishers, Dordrecht, 179–192, 1991.

14. Benacchio, L. and Nanni, M., "DIRA2: A database of astronomical catalogues", in *Astronomy from Large Databases II*, A. Heck and F. Murtagh, eds., European Southern Observatory, Garching, 201–206, 1992.

15. Benvenuti, P., "Evolution in the scientific uses of archives for space missions", in *Astronomy from Large Databases II*, A. Heck and F. Murtagh, eds., European Southern Observatory, Garching, 43–47, 1992.

16. Berners-Lee, T., "Electronic publishing and visions of hypertext", *Physics World*, **5**, 14–16, 1992.

17. Bogaschewsky, R., "Hypertext-/Hypermedia-Systeme – Ein Überblick", *Informatik Spektrum*, **15**, 127–143, 1992.

18. Brackeen, D., Grits, D., Mackraz, B. and Tognazzini, B., *Apple Human Interface Checklist*, 1989.

19. Carpenter, B. and Williams, D., "The importance of being networked", *Physics World*, **5**, 31–34, 1992.

20. Chambers, J.M., "Discussion of software environments for statistical computing", *Statistical Software Newsletter*, **15**, 81–84, 1989.

21. Chang, S.-K., Yan, C.W., Dimitrof, D. and Arnd, T., "An intelligent image database system", *IEEE Transactions on Software Engineering*, **14**, 681–688, 1988.

22. Cheng, Y. and Kashyap, R.L., "An axiomatic approach for combining evidence from a variety of sources", *Journal of Intelligent and Robotic Systems*, **1**, 17–33, 1988.

23. Chevy, D., "Internet information servers", Beloit College, 1992.

24. Chignell, M.H. and Parsaye, K., "Principles for applying intelligent databases", *Artificial Intelligence Expert*, 34–41, Oct. 1991.

25. Christian, C.A., Dobson, C. and Malina, R., "The EUVE data archives", in *Astronomy from Large Databases II*, A. Heck and F. Murtagh, eds., European Southern Observatory, Garching, 225–230, 1992.

26. Corbin, B.G., "Preserving the past: archival preservation of observatory publications via an optical disk project", in *Mapping the Sky*, 115–118, 1988.

27. Cornish, M., "The right tool for the job", *Computer Language*, **5**, 55–61, 1988.

28. Cox, R.T., "Probability, frequency and reasonable expectation", *American Journal of Physics*, **14**, 1–13, 1946.

29. Crawford, D., "Editorial pointers", *Communications of the ACM*, **35**, 2, 1992.

30. Date, C.J., *The SQL Standard*, Addison-Wesley, Reading, MA, 1987.

31. Davenhall, A.C., "Database applications in Starlink", in *Databases and On-Line Data in Astronomy*, M.A. Albrecht and D. Egret, eds., Kluwer Academic Publishers, Dordrecht, 165–178, 1991.

32. Davenhall, A.C., "Are commercial RDBMS suitable for manipulating astronomical catalogues?", in *Astronomy from Large Databases II*, A. Heck and F. Murtagh, eds., European Southern Observatory, Garching, 243–248, 1992.

33. Denning, P., "Electronic junk", *Communications of the ACM*, 163–165, March 1982.

34. Egret, D., Wenger, M. and Dubois, P., "The SIMBAD astronomical database", in *Databases and On-Line Data in Astronomy*, M.A. Albrecht and D. Egret, eds., Kluwer Academic Publishers, Dordrecht, 79–88, 1991.

35. Giommi, P. and Tagliaferri, G., *XIMAGE – an X-ray astronomy image analysis facility*, EXOSAT Observatory, Astrophysics Division, Space Science Department of ESA, ESTEC, NL-2200 AG Noordwijk, 1990.

36. Hafner, A., "dBase unter AutoLISP", *AutoCAD Magazin*, **1**, 76–78, 1992.

37. Hájek, P., "Combining functions for certainty degrees in consulting systems", *International Journal of Man-Machine Studies*, **22**, 59–76, 1985.

38. Hanisch, R.J., "STSDAS: The Space Telescope Science Data Analysis System", in *Data Analysis in Astronomy III*, V. Di Gesù et al., eds., Plenum Press, New York, 129–140, 1989.

39. Hanisch, R.J., "STSDAS: The Space Telescope Science Data Analysis System", in *Data Analysis in Astronomy IV*, V. Di Gesù et al., eds., Plenum Press, New York, 97–101, 1991.

40. Hanisch, R.J., "Network resources for astronomers", Space Telescope Science Institute, Baltimore, net_resources.mem, 1992a.

41. Hanisch, R.J., "Services available on the network", *Newsletter of the Americal Astronomical Society, special insert "Electronic Publishing in Astronomy; Projects and Plans of the AAS*, 12–13, 1992b.

42. Helou, G., Madore, B.F., Schmitz, M., Bocay, M.D., Wu, X. and Bennett, J., "The NASA/IPAC Extragalactic Database", in *Databases and On-Line Data in Astronomy*, M.A. Albrecht and D. Egret, eds., Kluwer Academic Publishers, Dordrecht, 89–106, 1991.

43. Jarke, M. and Vassiliou, Y., "A framework for choosing a database query language", *ACM Computing Surveys*, **17**, 313–340, 1985.

44. Johnson, D., "On holy wars and a plea for peace: MMST language s election", Texas Instruments, 16 July 1990, 1990.

45. Johnston, M.D., "StarView: The astronomer's interface to the Hubble Space Telescope archive", in *Astronomy from Large Databases II*, A. Heck and F. Murtagh, eds., European Southern Observatory, Garching, 75–83, 1992.

46. Krol, E., *The Whole INTERNET – User's Guide & Catalog*, O'Reilly & Associates, Inc., Sebastopol, CA, 1992.

47. Landsman, W.B., *The UIT Database System*, ST Systems Co., 1992.

48. Lubinsky, D.J., "Comment: two functional programming environments for statistics – Lisp-Stat and S", *Statistical Science*, **6**, 1991.

49. Lucy, L.B., "An iterative technique for the rectification of observed distributions", *Astronomical Journal*, **79**, 745–754, 1974.

50. Marcus, A. and van Dam, A., "User-interface developments for the nineties", *IEEE Computer*, **24**, 49–57, 1991.

51. Michalewicz, Z., *Statistical and Scientific Databases*, Ellis Horwood, New York, 1991.

52. Murray, S.S., Brugel, E.W., Eichhorn, G., Ferris, A., Good, J.C., Kurtz, M.J., Nousek, J.A. and Stoner, J.L., Astrophysics Data System (ADS), in *Astronomy from Large Databases II*, A. Heck and F. Murtagh, eds., European Southern Observatory, Garching, 387–391, 1992.

53. Murtagh, F., "Large databases in astronomy", *Encyclopedia of Computer Science and Technology*, **21**, Suppl. 6, 205–213, 1990.

54. Nielsen, J., "The Art of navigating through hypertext", *Communications of the ACM*, **33**, 296–310, 1990.

55. Norvig, P., *Paradigms of Artificial Intelligence Programming: Case Studies in Common Lisp*, Morgan Kaufmann, 1991.

56. Ochsenbein, F., "The ESO archive project", in *Databases and On-Line Data in Astronomy*, M.A. Albrecht and D. Egret, eds., Kluwer Academic Publishers, Dordrecht, 107–114, 1991.

57. Ozsoyoglu, G. and Ozsoyoglu, Z.M., "Statistical database query languages", *IEEE Transactions on Software Engineering*, **SE-11**, 1071–1081, 1985.

58. Page, C.G., "Role of the relational database in astronomical data reduction", in *Astronomy from Large Databases II*, A. Heck and F. Murtagh, eds., European Southern Observatory, Garching, 411–415, 1992.

59. Parsaye, K., Chignell, M., Khoshafian, S. and Wong, H., "Intelligent databases", *Artificial Intelligence Magazine*, **5**, 38–47, 1990.

60. Pasian, F. and Richmond, A., "User interfaces in astronomy", in *Databases and On-Line Data in Astronomy*, M.A. Albrecht and D. Egret, eds., Kluwer Academic Publishers, Dordrecht, 235–252, 1991.

61. Péron, M., Ochsenbein, F. and Grosbøl, P., "The ESO-MIDAS table file system", in *Astronomy from Large Databases II*, A. Heck and F. Murtagh, eds., European Southern Observatory, Garching, 433–438, 1992.

62. Pirenne, B., Albrecht, M.A., Durand, D. and Gaudet, S., "STARCAT: an interface to astronomical databases", in *Astronomy from Large Databases II*, A. Heck and F. Murtagh, eds., European Southern Observatory, Garching, 447–453, 1992.

63. Pomphrey, R. and Good, J., "The Astrophysics Data System – an overview", *Information Systems Newsletter*, 39–42, May 1990.

64. Ponz, D. and Murtagh, F., "MIDAS TABLES: present status and future evolution", in *Astronomy from Large Databases: Scientific Objectives and Methodological Approaches*, F. Murtagh and A. Heck, eds., European Southern Observatory, Garching, 441–446, 1987.

65. Press, L., "Collective Dynabases", *Communications of the ACM*, **35**, 26–32, 1992a.

66. Press, L., "The Net: progress and opportunity", *Communications of the ACM*, **35**, 21–25, 1992b.

67. Raimond, E., "Archives of the Isaac Newton Group, La Palma and Westerbork observatories:, in *Databases and On-Line Data in Astronomy*, M.A. Albrecht and D. Egret, eds., Kluwer Academic Publishers, Dordrecht, 115–124, 1991.

68. Research Systems, I., *IDL User's Guide – Interactive Data Language, Version 2.2, August 1991*, Research Systems Inc., 777 29th Street, Suite 302, Boulder, CO 80303, 1991.

69. Rey-Watson, J.M., "Access to astronomical literature through commercial databases", in *Astronomy from Large Databases: Scientific Objectives and Methodological Approaches*, F. Murtagh and A. Heck, eds., European Southern Observatory, Garching, 453–458, 1987.

70. Richardson, B.H., "Bayesian-based iterative method of image restoration", *Journal of the Optical Society of America*, **62**, 55–59, 1972.

71. Richmond, A., McGlynn, T., Ochsenbein, F., Romelfanger, F. and Russo, G., "The design of a large astronomical database system", in *Astronomy from Large Databases: Scientific Objectives and Methodological Approaches*, F. Murtagh and A. Heck, eds., Garching, European Southern Observatory, 465–472, 1987.

72. Rosenthal, D.A., "Applying artificial intelligence to astronomical databases – a survey of applicable technology", in *Astronomy from Large Databases: Scientific Objectives and Methodological Approaches*, F. Murtagh and A. Heck, eds., Garching, European Southern Observatory, 245–258, 1987.

73. Rothrock, M.E., "The power of LISP for C programmers", *Artificial Intelligence Expert*, **4**, 15–21, 1990.

74. Rumble, J.R. and Smith, F.J., *Database Systems in Science and Engineering*, Adam Hilger, Bristol, 1991.

75. Russo, G., Richmond, A. and Albrecht, R., "The European scientific data archive for the Hubble Space Telescope", in *Data Analysis in Astronomy II*, V. Di Gesù et al., eds., Plenum Press, New York, 193–200, 1986.

76. Sajeev, A.S.M. and Hurst, A.J., "Programming persistence in chi", *IEEE Computer*, **25**, 57–66, 1992.

77. Schreier, E., Benvenuti, P. and Pasian, F., "Data archive systems for the Hubble Space Telescope", in *Databases and On-Line Data in Astronomy*, M.A. Albrecht and D. Egret, eds., Kluwer Academic Publishers, Dordrecht, 47–58, 1991.

78. Semmel, R.D., "An overview of automated query formulation", Johns Hopkins University, Baltimore, RMI-91-002, 1991.

79. Shepp, L.A. and Vardi, Y., "Maximum-likelihood reconstruction for emission tomography", *IEEE Transactions on Medical Imaging*, **MI-1**, 113–121, 1982.

80. Shoshani, A. and Wong, H.K.T., "Statistical and scientific database issues", *IEEE Transactions on Software Engineering*, **SE-11**, 1040–1047, 1985.

81. Sipelstein, J.M. and Blelloch, G.E., "Collection-oriented languages", *Proceedings of the IEEE*, **79**, 504–523, 1990.

82. Stallman, R., *GNU Emacs Manual*, Free Software Foundation, Cambridge, MA, 1987.

83. Steele, G.L., *Common Lisp – The Language*, 2nd ed., Digital Press, 1990.

84. Tarasko, M.Z., Obninsk, FEI-156, 1969.

85. Tesler, L., "Foreword", in *Dylan – an object-oriented dynamic language*, A.M. Shalit, J. Piazza and D. Moon, eds., Apple Computer, Eastern Research and Technology, Cambridge, MA, 7–8, 1992.

86. Tierney, L., XLISP-STAT, a statistical environment based on the XLISP language, Technical Report No. 528, University of Minnesota, School of Statistics, 1988.

87. Tierney, L., *Lisp-Stat: An Object-Oriented Environment for Statistical Computing and Dynamic Graphics*, New York, Wiley, 1991.

88. Touretzky, D.S., "How Lisp has changed", *Byte*, 229–234, Feb. 1988.

89. Van Steenberg, M.E., Gass, J., Brotzman, L., Warnock, A., Kovalsky, D. and Giovane, F., "STELAR: An experiment in the electronic distribution of astronomical literature", *Newsletter of the Americal Astronomical Society, special insert "Electronic Publishing in Astronomy; Projects and Plans of the AAS"*, 11, 1992.

90. Warnock, A., Gass, J., Brotzman, L., Van Steenberg, M.E., Kovalsky, D. and Giovane, F., "On-line WAIS search capability brings astronomy to the Internet", *Newsletter of the Americal Astronomical Society, special insert "Electronic Publishing in Astronomy; Projects and Plans of the AAS"*, 10, 1992.

91. Watson, J.M., "Astronomical bibliography from commercial databases", in *Databases and On-Line Data in Astronomy*, M.A. Albrecht and D. Egret, eds., Kluwer Academic Publishers, Dordrecht, 199–210, 1991.

92. Weerawarana, S. and Wang, P.S., "A portable code generator for CRAY Fortran", *ACM Transactions on Mathematical Software*, **18**, 241–255, 1992.

93. Weiss, J.R. and Good, J.C., "The NASA Astrophysics Data System", in *Databases & On-line Data in Astronomy*, M.A. Albrecht and D. Egret, eds., Kluwer Academic Publisher, Dordrecht, 139–150, 1991.

94. White, N.E. and Giommi, P., "The EXOSAT database system", in *Databases and On-Line Data in Astronomy*, M.A. Albrecht and D. Egret, eds., Kluwer Academic Publishers, Dordrecht, 11–16, 1991.

95. Winston, P.H. and Horn, B.K.P., *Lisp*, 3rd Ed., Addison-Wesley, Reading, MA, 1989.

96. Zimmermann, H.U. and Harris, A.W., "Data from the ROSAT archive", in *Databases and On-Line Data in Astronomy*, M.A. Albrecht and D. Egret, eds., Kluwer Academic Publishers, Dordrecht, 1–9, 1991.

Appendix 1: Programming Languages

We are currently most likely witnessing a starting transition from the prevailing "single-language only" programming paradigm, which for some time had replaced the earlier Assembler/Fortran language tandem, towards another multi-lingual era, where Fortran and/or C act as low-to-medium level, efficient programming work-horse(s) to be complemented with another, yet-to-be-agreed-upon high-level programming language suitable for cross-architectural scientific programming in the 1990s. Such a (hopefully widely accepted) language should be practical on small machines (e.g. for development), offer static language users an attractive dynamic alternative supporting rapid prototyping and – a sine qua non – achieve acceptable run-time performance in production code (cf. Tesler, 1992).

Appendix 2: Lisp

Common Lisp is a standardized, very high-level, functionally-oriented, dynamic, general purpose programming language including the first standardized object system, CLOS. A de facto standard for graphics, CLIM, is currently emerging. Common Lisp is available on practically all hardware platforms in major use, and there are several public domain implementations of the language. Common Lisp code can be interpreted, but is usually incrementally compiled for higher efficiency.

Common Lisp has a particularly rich and complete set of operators for strings – an important asset in information retrieval. The set includes operators which for a string determine its length (i.e. the number of characters), fill it with some value, find or return a substring, replace a substring by something else or reverse it. Furthermore there are operators for two strings, including concatenation, comparison and reduction. Most of the operators for strings equally well operate on lists, the fundamental data structure of Lisp.

Further information on Common Lisp can be found in (Cornish, 1988; Touretzky, 1988; Winston and Horn, 1989; Johnson, 1990; Rothrock, 1990; Steele, 1990; Adorf, 1991; Norvig, 1991).

4.6.1 Appendix 3: Programming a Spreadsheet

Apart from holding a value, each cell of a spreadsheet may also hold a mini-program, namely the code fragment necessary to re-compute the cell's value from the values of other cells in the spreadsheet. The code associated with a cell is normally hidden from view and only the data value is displayed in the cell.

From a software engineering point of view, a spreadsheet can be considered an object-oriented system with functional, equation-based rules. (Rule-based systems have become prominent in the expert systems area.) The basic object is the cell which stores the cell's name, its value, its display format, its associated code, and an optional textual note. Defaults are provided by the system. The user is usually unaware of the fact that (s)he is using a programmable, object-oriented, rule-based system, since the spreadsheet paradigms match the tasks at hand in such a natural way.

Each code fragment can be considered as a data-driven, forward-chaining rule disguised as a function equation of the generic form

```
cell_value = function(other_cell_values)
```

where the left hand side of the equation, being redundant, is suppressed. (Each code fragment is associated with a specific cell.) The function can be either intrinsic or user-defined. The intrinsic functions in MS Excel, for instance, comprise roughly those built into Fortran plus several others typical for business applications. All program constructs return values, including IF-statements which are of the form

```
if(condition, then_clause, else_clause)
```

Functions can be recursively nested (up to several levels deep).

Whenever a value in the function on the right hand side of an equation changes the rule is automatically marked as fireable. The order of the rule evaluation is arbitrary and controlled by the spreadsheet, not by the user. Upon rule firing the function on the right hand side will be re-evaluated and the result stored in the cell.

Cell reference is relative by default. For instance, if one wants to "add two columns", A and B say, to form a third, C say, then it is sufficient to write the code fragment (= rule)

```
= A1 + B1
```

into cell C1. The cells C2, C3, ... below C1 are then programmed by "spreading" the code fragment along the column using a mouse dragging operation. As a result cell C2 receives code fragment "= A2 + B2", cell C3 fragment "= A3 + B3" etc. An absolute reference to a specific column or a specific row can be enforced, if required.

Circular rules, i.e. rules that depend on each other, are usually prohibited as a safeguard measure, but may be enabled for special purposes such as fixed-point iterations.

The code fragments, of which a spreadsheet program usually consists, can be augmented by user-defined functions collected in separate "macro sheets". The set of intrinsic functions that can be used is identical to those invocable from the spreadsheet menus.

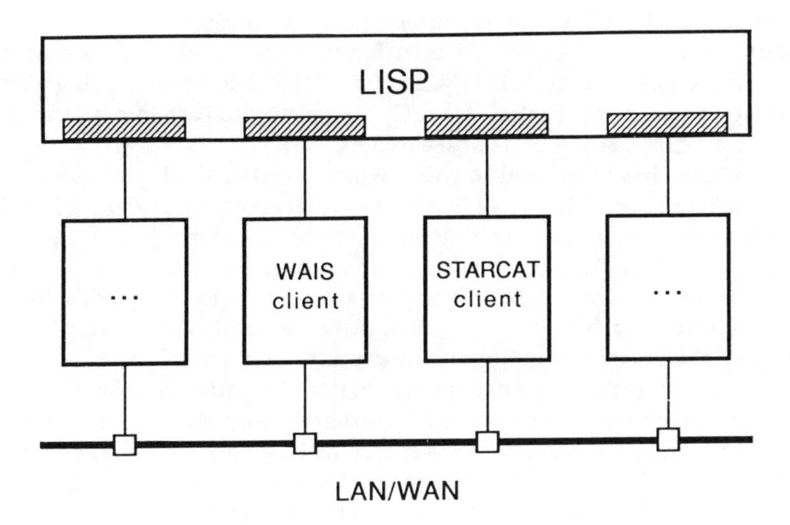

Figure 4.7: Architecture of an experimental Lisp top-level integrating a WAIS- and a STARCAT-client into a single homogeneous environment.

Appendix 4: Merging WAIS and STARCAT in Lisp

In order to further explore the idea of a common interactive and programmable top-level for information retrieval (Figure 4.7), an experimental WAIS-client has been implemented in Common Lisp which operates by running a standard WAIS-client in a (Unix) child process. The results of a WAIS search are fed back through a stream directly into Lisp.

A user interacts with the WAIS/LISP client via two top-level functions, wais-search and wais-retrieve. The first of these accepts a query and one or more WAIS sources and returns a list of document specification. The second accepts a document specification and retrieves a document directly into a Lisp data-structure where it can be processed further, if required.

WAIS document specifications and queries are always coded as Lisp structures, so it is almost trivial to write Lisp accessor functions for decoding them. For instance, suppose that number-of-lines is an accessor function retrieving the number of lines from a document specification bound to the variable `doc-specs`, say. Then the one-liner

```
(sort doc-specs '> :key 'number-of-lines)
```

returns a list of document specifiers sorted according to the descending size of the document. It is also easy to select various subsets of targets.

Lisp functions can also be "mapped". For instance, the line

```
(map 'vector 'number-of-lines doc-specs)
```

extracts from a sequence of document specifiers bound to the variable `doc-specs` a vector with the number of lines of the corresponding documents.

Similarly, a simple STARCAT/LISP-client has been crafted which allows the retrieval of a set of HST Catalog records directly into Lisp. The ability of importing a subset of the information stored in a central, remotely accessible database into a local database facilitates a statistical analysis of database records.

The Lisp-client has been used to check which targets have been observed by the Hubble Space Telescope. In January 1993 the HST Catalog contained about 2,500 different target designations, many of which, however, were synonyms pointing to the same physical object. The star Ap Lib, for instance, was recorded as `APLIB` and `AP-LIB`. There were more than half a dozen different target names for Pluto, not all of which would match a wildcarded search for `*PLUTO*`. Summary information of this kind is vital both for debugging a database and for formulating successful queries.

Using Common Lisp's rich set of built-in operators it is particular easy to form subsets of records and to combine these subsets in an arbitrary way. Also the merging of information from different information sources, such as WAIS and STARCAT, is facilitated. Carrying the idea a bit further one can imagine sophisticated cross-database searches where the result from querying one database controls the search in another.

Chapter 6

What Hypertext can do for Information Retrieval

Ronald Bogaschewsky and Uwe Hoppe

Institut für Betriebswirtschaftliche Produktions- und Investitionsforschung
Georg-August-Universität Göttingen
Platz der Göttinger Sieben 3
DW-3400 Göttingen (Germany)
Email: boga@dgowiso1.bitnet

Summary

Conventional information retrieval systems (IRS) do not take into account the broad range of information needs different users have at various times. Hybrid systems combining the functionalities of conventional IRS and hypertext systems are a promising approach for building user-oriented systems for effective and efficient information retrieval. After the concept of hypertext and the functionality of hypertext systems is outlined, we discuss how information retrieval and hypertext can be integrated. The specific organization of the information base and the different ways to access the information are analyzed. Furthermore, information retrieval models are outlined and dedicated retrieval models for hypertext are presented.

6.1 The Information Problem and Computer Support

Regarding the development of human knowledge in the past decades and especially the last years an explosion is evident. Since new knowledge is been written down in some way, the knowledge explosion is correlated with an exponential growth of published articles, books, reports, etc. Furthermore, lots of (mostly textual) non-scientific data, e.g. novels, operator's manuals for household equipment, entertainment literature, etc.,

A. Heck and F. Murtagh (eds.), Intelligent Information Retrieval: The Case of Astronomy and Related Space Sciences, 81–102.

constitute a tremendous amount of potential information sources. The main problem for a person who wants to inform her/himself about a specific subject is nowadays – at least in developed countries – how to extract the right information from these not overseeable masses and not to find information at all. Even for non-scientists the only way to cope with this problem is to ask for help from a computer-based system that stores the information itself or at least representatives of the actual information sources. This system should provide quick access to the desired information or should act as an information source by guiding the user through the database or prompting her/him to explore what kind of information is stored in the system. In this sense a system may help the user on the one hand to get started when s/he is looking for some information source that physically exists, e.g. a book, and that s/he wants to study in the conventional manner, e.g. by reading it. On the other hand a system may actually augment and amplify the users intellect by providing additional functionality besides reading a book in a sequential manner.

How people typically act when looking for information in a more serious way, which does not mean a "couch potato" scanning the TV program of the week, can be described by a model developed by Ellis (1989), who identified six categories that represent activities researchers in the social sciences are engaged in when looking for information (see also Waterworth and Chignell, 1991):

1. *starting*: initial search; identifying material which matches terms of subject descriptions
2. *chaining*: following chains of citations or other referential connections
3. *browsing*: examining a set of references from the area of potential interest
4. *differentiating*: filtering the examined sources for further use applying self-defined criteria
5. *monitoring*: keeping track of developments in a specific domain, e.g. monitoring specific periodicals or publications of certain authors
6. *extracting*: systematically working through particular sources.

Even though these criteria may not be valid for any specific type of user in every possible situation s/he is looking for information, this approach demonstrates the complexity of the process of information seeking.

Conventional systems that support activities in the above described process are *information retrieval systems* (IRS). These systems are concerned with the representation, storage, organization, and accessing of information units – usually referred to as documents – or their representatives (see Salton and McGill, 1983). Most systems that are in use in reality only store representations of the actual information sources. These representations typically consist of bibliographic data like author, year, etc., information where to find the physical sources, e.g. signatures from libraries, and rather rudimentary contextual data provided by the title and possibly subtitles, a short abstract, or keywords.

Therefore, these systems can be classified as *document or reference retrieval systems* (see Steinacker, 1975; Bloech and Bogaschewsky, 1986).

Systems that provide a larger part of or even the entire information source can be classified as *fulltext retrieval systems*. However, the term *information retrieval system* usually covers both system types, since they cannot be arbitrarily classified in their everyday use. More often conventional IRS as mentioned first are differentiated from the fulltext RS.

Obviously, conventional IRS that basically give only information about the existence (or non-existence) of documents and their whereabouts cannot effectively support all six phases that were outlined above. They may help the user in getting started (initial search), give some assistance in the differentiation activities (filter documents) and support the user in keeping track of developments (monitoring), insofar as they were edited into the database. Conventional IRS give assistance by providing a user interface that allows search requests formulated by the end-user or a (human) intermediary. The system software then searches[1] for document representations that match this request and presents a list of "hits". The user can view this list and eventually redefine her/his request in an iterative manner, until s/he decides to stop this process and tries to get the actual (physical) sources according to the final list of retrieved documents.

Further support is provided by a fulltext retrieval system by giving the user online access to the actual information source in the computer. Therefore, s/he can further differentiate (activity 4) between more and less useful sources by taking a deeper look into their contents, s/he may extract information (activity 6) by working through the sources and jot down citations (chaining – activity 2) s/he finds in the documents for further investigation. And last but not least s/he is able to browse through the source (activity 3) in the same (basically sequential) way s/he would do if the physical source would be available. The advantages of such a computer-based system are first of all the faster access to the sources, their availability to several users at a time, their completeness and correctness (hopefully), and possibly the existence of some additional features for storing, sorting, extracting, etc. A lot of systems provide (electronic) indexes or thesauri in order to give the user initial orientation about the contents of the database and to assist her/him in formulating search requests using the allowed vocabulary. However, the system only emulates activities the user otherwise would do her/himself, e.g. going to the library, checking out the sources and working through (reading) them. Even the electronic keyword lists may be used in the conventional paper-based form.

Using the full-blown capabilities of computers would mean to add features the user would not be able to accomplish without computer-based help. From the six phases of the information seeking process outlined above, several requirements for a computer-based support system can be derived. These requirements can be summarized that the user should be able to move freely through the information space consisting of the documents stored in the database. S/he should be able to access documents that are related to each other in some way, to extract parts of documents, to store and structure them, and so on. An approach that is well suited to satisfy these requirements is the hypertext concept

---

[1]In order to search through the database effectively and efficiently the system software applies specific string searching algorithms (see Baeza-Yates, 1992; Gonnet and Baeza-Yates, 1991; Baeza-Yates and Gonnet, 1992; Wu and Manber, 1992).

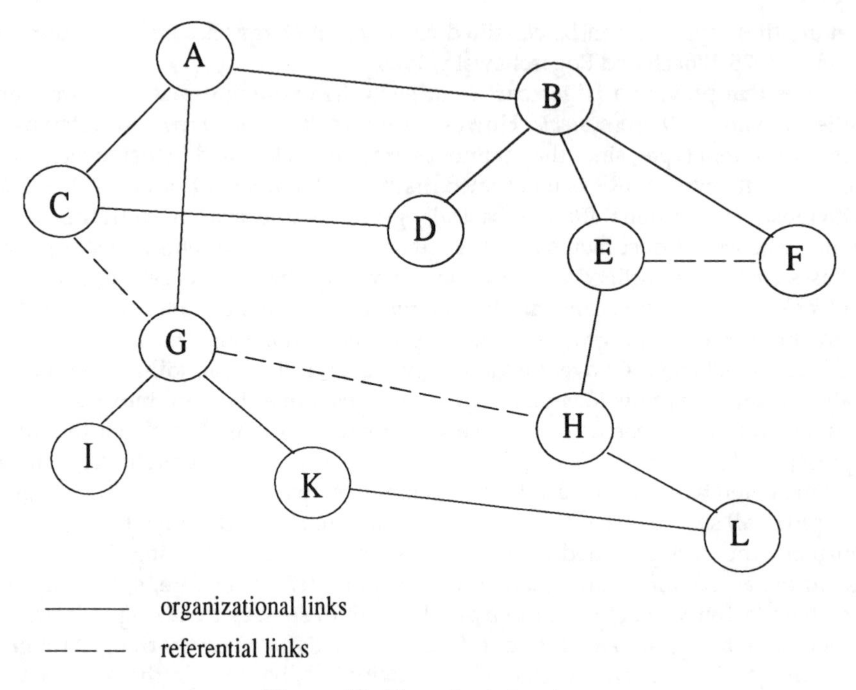

Figure 6.1: Hypertext structure.

described in the next section. However, some difficulties have to be solved in order to build real-life hypertext-based systems for effective and efficient information retrieval. Some approaches for using hypertext for information retrieval are described in section 6.3.

6.2 Hypertext: an Information Network

6.2.1 The basic concept of hypertext

A hypertext consists of *information units* – often called *nodes* – connected by *links*. The logical model of a hypertext represents a graph or network of nodes connected by links. Due to the given freedom in linking nodes a hypertext can represent any structure such as linear, hierarchical, or netlike. Figure 6.1 shows a typical hypertext structure.

The user of a computer-based hypertext system does not see this logical structure, except if the system provides her/him with a *graphical browser* that makes it visible. The information units are presented to the user either in parts or as a whole on the screen or in windows. Full screen representations allow the user to see one unit at a time. If the unit fills up more space than the screen is able to show, the system either lets the user scroll up and down or allows to toggle from one screen to the next/previous. Systems

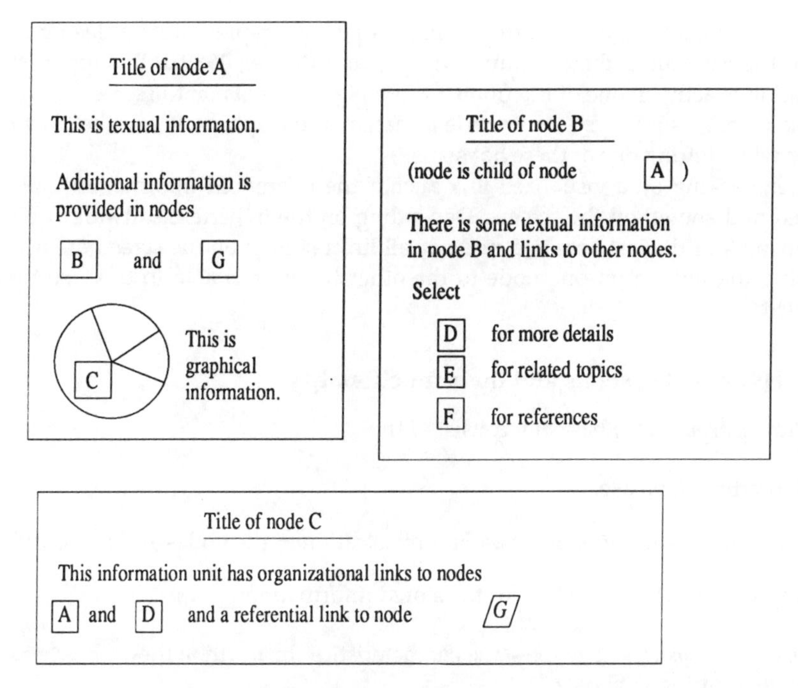

Figure 6.2: Visualization of information units.

supporting windowing techniques provide this functionality within each window the user can open at a time. Figure 6.2 shows the contents of three nodes in three different windows on the screen.

The visualized nodes – often called *cards* – in this example are named A, B, and C. These names could also be used as the *label* or *descriptor* of the node. The nodes in Figure 6.2 show textual and graphical information. In case the unit contains multimedia information (voice, video sequences, etc.) appropriate presentation techniques and hardware have to be installed. The contents of the nodes differentiate (plain) hypertext from multimedia hypertext, or *hypermedia*.

The nodes in Figure 6.2 also show letters enclosed in rectangles that are used in this example to identify *anchors of links* to other nodes. From node A links exist to nodes B, C, and G, from node B to nodes A, D, E, F, and so on. In this case the beginning of a link is located in a specific part of a node and the destination is an entire other node. Depending on the functionality of the hypertext software it can also be possible to allow links to parts of other nodes or just from a node as a unit to another node.

The way link anchors can be visually represented is either defined by the hypertext software or can be designed by the user. Most common is the usage of highlighting text, printing text in italics or bold, and adding *buttons* (arrows, circles, boxes) or more complex graphical symbols, such as folder, paper basket, etc. Graphical symbols are known as *icons*, a term derived from religious pictures representing a deeper meaning used in the Russian Orthodox church (Hayes and Pepper, 1989). To support the user in pointing exactly on one of the (more or less big) hot spots on the screen representing the link anchors, some systems provide a change in the appearance of the cursor or the pointer when hitting one of these hot spots.

When clicking on a visualized link anchor the referenced information unit will be accessed and shown on the screen. Depending on the hypertext software used, either the hot spots of the last accessed node or all links shown on the screen are active. By following the links from one node to the other the user is able to traverse the entire hypertext.

6.2.2 Hypertext systems and their functionality

Hypertext systems (HTS) basically consist of the

- hypertext database,
- software to support storage and administration of the nodes and links, and
- input and output facilities for the stored information.

Hypermedia systems have the same characteristics. In addition they are able to handle and present multimedia data.

Different approaches for implementing the database have been chosen, such as using relational database systems, object-oriented database systems, or enhanced text editors. The input facilities are usually text editors, graphical editors, or combined text/graphical editors. For sound/voice and video input specific hardware extension boards are needed or have already been integrated in so-called multimedia machines. For multimedia output the workstations need speakers connected to the voice card, specific hardware processing graphical data, and high resolution screens. Available HTS differ widely in the functionality they offer and how they are implemented. Only recently, efforts to develop standards have been started.[2]

A rather larger difference can be seen in the functionality of hypertext systems concerning the definition of different node and link types and in the ways the hypertext and the nodes can be structured. When developing a hypertext database, a lot of questions have to be answered. What are the information needs of the users? Should there be different ways to access information depending on the knowledge of the user in the domain? How should the domain be structured? The strong flexibility of (nonlinear) hypertext often causes a stronger mental loading to the developer which Conklin (1987, p. 40) calls *"cognitive overhead"*, compared to the development of linear documents.

[2]Groups working on standards as a foundation for interchange from/to HTS and other capabilities of HTS are the HyTime Group and the Davenport Group. HyTime stands for Hypermedia/Time-based Document Structuring Language and the group proposed an International Standard balloted as a Committee Draft (ISO/IEC CD 10744; Newcomb et al., 1991). The Davenport Group is developing a common interchange format based on the Structured General Markup Language (SGML; SIGLINK, 1992).

When implementing a hypertext as an information system the relevant domain has to be preselected. Then it has to be decided if the hypertext should focus on a somewhat specific and narrow domain, or if it should also provide basic information in the domain to help the user understand more complicated contexts. In order to satisfy the different information needs of the users, it seems to be reasonable to provide access to the information in an extremely flexible manner. Beginners and users who seldom access the hypertext may have to be guided through the information network.

The definition of the nodes of a hypertext is very important, because they represent the basic information units. The linking of the nodes gives the hypertext its structure and determines pathways within the network. Therefore, the suitability of a system is determined by its functionality concerning the freedom to define nodes and links in a way that is suitable for implementing the information of the domain for different user types.

Creating *typed* nodes by assigning a label or descriptor to them helps in structuring the hypertext. For the same reason some systems provide the feature of *clustering* nodes. A cluster can be seen as a meta-node that can be linked to other (meta-)nodes giving a structure at a more aggregated level. *Semi-structured* nodes can have named and/or formatted fields. In this way nodes could be used for example to provide bibliographic and contextual information about a monograph or an essay. When building up the information network by linking nodes to each other, basically two types of links can be used. *Organizational links* are used to implement a specific structure, such as hierarchies and trees. *Referential links* connect nodes to each other that may address the same topic, or they may be used to link additional information about a topic or bibliographical references to a node, etc. Pretty often hypertext systems do not differentiate between these types of links, and it depends on the complexity of the hypertext and the needs of its developer, if this functionality is needed. Both link types may be *directional*, providing a link from the source to the destination, or *bi-directional* providing a link between the nodes in both directions (see Figure 6.1).

Another link type sometimes provided by the system software is the keyword link (Conklin, 1987, p.35). It is used when nodes can be accessed via keyword search. Since keywords can be assigned to more than one node, a keyword link is a sort of clustered link. Clustered links can be used to reference more than one node at a time. More flexible systems allow *user-defined links* for relative complex structures. In this way special types of semantic networks could be realized. More specific link types are provided by problem exploration tools (see below), e.g. issue-based information systems, that are of no concern in this essay. Another rather specific feature is the possibility to call *procedures* that compute or extract data from some source and eventually modify nodes and/or links in the hypertext when activating a link. By implementing this functionality, a hypertext becomes a dynamic system compared to the static structure of a database.

6.2.3 Orientation and navigation in hypertext

As mentioned above, a user typically traverses a hypertext by following links. A problem that occurs when traversing a hypertext is orientation. Compared to linear structures such as printed text, the nonlinearity of hypertext is very uncommon for users, sometimes resulting in disorientation. This means, that users loose their sense of location and direction in the information network. Therefore, the system has to provide assistance in order to avoid that the user gets *"lost in hyperspace"*.

Various concepts have been developed to facilitate navigation and orientation within a hypertext. One of the most helpful features is the *graphical browser*. Graphical browsers offer a view of the logical structure of the hypertext with its nodes and links as shown in Figure 6.1. By labeling the nodes the user can find where his/her current location is in the network and where s/he can progress from here. Depending on the system software, either a (passive) map provides an overview or an active map can be used to access nodes directly by clicking on one of the nodes shown. Graphical browsers have been developed where the user can decide how much of the network and what type of links s/he wants to see at a time, e.g. fisheye-views (see Hofmann, 1991).

Other features to help the user in orientating and navigating are ordered lists of already "visited" nodes and of most often accessed nodes, buttons for "jumping" to a starting or home node, to the beginning of "chapters", or to user-marked nodes. The developer of the hypertext has the opportunity to implement *"guided tours"* which users can follow. Depending on his/her information needs and domain knowledge, the user may be able to choose between different guided tours.

6.2.4 A short history of hypertext

The basic concept of hypertext was outlined for the first time as early as 1945. Vannevar Bush presented the concept of a "memex" which was meant to be "...a device in which an individual stores his books, records, and communications, and which is mechanized so that it may be consulted with exceeding speed and flexibility. It is an enlarged intimate to his memory" (Bush, 1945, p. 106ff.). The memex was intended to link information units without any restrictions. More or less every scientific publication all over the world was aimed to be stored in this information network (Fraase, 1990). Due to the state-of-the-art of storage techniques in the forties, Bush suggested photocells and microfilm as storage media (Conklin, 1987).

The first computer-based system following Bush's ideas was H-LAM/T (Human using Language, Artifacts, and Methodology, in which he is Trained) developed by Douglas Engelbart (1963) at Stanford Research Institute. The system was meant to dynamically change components with the user in a symbiosis which had the effect of "amplifying" the native intelligence of the user (Conklin, 1987, p. 22). Engelbart's principle of hypertext as "A Conceptual Framework of Man's Intellect" (Engelbart, 1963) was implemented in 1968 as NLS (oN-Line System), which later was further developed and commercially distributed as *Augment*. Engelbart also pioneered some features which are now widely used in modern user interfaces, such as the *mouse* as a pointing device.

Ted Nelson (1967) designed the system *Xanadu*, a "...repository publishing network for anybody's documents and contents, which users may combine and link to freely" (Nelson, 1988). Xanadu was designed as a worldwide network based on logical links between information units that are physically distributed on different machines (nodes).

There is no limit to the number of nodes. To identify documents that are stored within the nodes, Xanadu uses so-called Humbers (Humongous Numbers) that can represent up to 10^{300} units (Nelson, 1988). Users can declare documents – which can contain textual, graphical, and vocal information – as either public or private. Ten to twenty percent of the data transmission costs paid by the user are credited as royalties to the provider of the document. The commercial version of Xanadu is supposed to be available under Unix (Sun and Apple) and OS/2 (Fraase, 1990).

It seems that in the beginning of its development, hypertext has been used as a tool for storing, administrating, and accessing documents. The hypertext approach does not really differentiate between the structuring of one big document or the access to a bulk of interconnected – possibly physically distributed – different documents. All the above mentioned systems can therefore be classified as *macro literary systems* (Conklin, 1987). Regarding the type of application, different hypertext systems are designed to be used for (Conklin, 1987) further differentiation into *problem exploration tools, browsing systems*, and *general hypertext technology*. Due to recent developments it seems reasonable to separate *authoring systems* and *interactive presentation systems* from the classes listed above (Bogaschewsky, 1992).

In discussing the aspects of information retrieval, first of all the macro literary systems are of concern. Furthermore, (general) hypertext systems that offer a large flexibility and functionality are a matter of interest, since they can be used to build literary systems. An overview of the specifics of some systems of the other classes mentioned is given by Bogaschewsky (1992); Carando (1989); Conklin (1987); Gloor (1990); Kuhlen (1991); Nielsen (1989).

6.3 Hypertext and Information Retrieval

6.3.1 What does hypertext for information retrieval look like?

When examining the opportunities of hypertext for information retrieval it becomes obvious that using a computer-supported system that is based on the hypertext concept is quite different to conventional retrieval systems. It should again be stated here that conventional IRS do not usually store the actual information source. They only provide the user with information about the existence of information sources, some bibliographic and rather rudimentary contextual data, and information where to find the physical sources whose representations were retrieved in the database. Since HTS are able to store the entire information source itself, they can be compared to fulltext retrieval systems rather than to reference retrieval system. The opportunity to – more or less – freely link information units to each other differentiates HTS from fulltext retrieval systems developed so far. The structure of the hypertext that represents the information base is not predefined. Depending on the intended usage and the anticipated users and user groups the hypertext may have different structures. The basic unit for conventional information retrieval is the document, representing a monograph, an essay in a periodical, etc. In fulltext retrieval systems this document can be stored and accessed in its entirety. In addition it is possible to structure the document itself when employing a hypertext system. The tailorability of the information sources makes the big difference between (reference or fulltext) retrieval systems and HTS. This means that information systems where the database is organized as a hypertext have, by far, greater potential than IRS. The entire database can be structured by linking documents and parts of the (well structured) documents to each other providing additional information compared to unstructured (e.g. sequential) databases.

So far, a lot of researchers who analyzed how information retrieval in hypertexts can be done assumed that a document is represented by exactly one node. This only makes sense if the (presumably short) documents, e.g. abstracts, comments, etc., should not have an internal structure. In this case hypertext differs from other fulltext retrieval system only by the opportunity of linking whole documents. The advantage of hypertext is still quite big, since the user has the additional feature to access related information by following the links. Of course, there is some extra (possibly considerable) effort to build up and maintain such a hypertext network[3] compared to a database consisting of isolated information units. Consequently, the pros and cons of hypertext and conventional systems for information retrieval depend on the user's needs and the acceptable cost and efforts in developing and maintaining such a system to satisfy these needs.

Hypertext systems develop their full-blown functionality for information retrieval when the documents themselves are structured. This means that the documents themselves consist of several nodes connected via links. A monograph may be subdivided into chapters, sections and pages resulting in a hierarchical structure. Additional referential links may connect contextual information located in different parts of the book. In special cases the entire network may consist of only one (hyper-)document, e.g. an encyclopedia, or a dictionary. Each entry may consist of one or even several nodes connected to other information units. The modularity of the nodes is a design issue which the author of the hyperdocument has to cope with.

When the network consists of several documents each of them represented by one or more nodes, the structure of the hypertext gets more complex. The nodes that represent a given document can be joined into a *cluster*. A cluster of different documents, a meta-document, may represent an anthology or all the books placed on a specific shelf in a library. Clustering several (meta-)documents may then represent an entire library. Sections of a given document can be clustered into chapters. Criteria for clustering different documents may be author, year, contents, etc. Nodes from various documents can be clustered depending on contents, type of media (text, graphics, video, sound), etc. Whether or not the hypertext should be subdivided into clusters, and the criteria to be applied, depends on the expected information needs. A cluster represents an aggregated information unit that has to have its own meaning or represents additional value for the user. It is a system design issue if nodes may be able to become part of different

[3]Research has been done on how to (semi-)automate the conversion from paper documents to hypertexts (Stotts and Furuta, 1990; Bernstein, 1990) and how to assist in building hypertext fragments using a knowledge base that is customized by the user (Clitherow et al., 1989). A hypertext-based IRS that runs successfully is Justus (Wilson, 1988; see also Ellis, 1990, p. 113ff.). Justus is based on the Unix version of the Hypertext System 'Guide' from OWL International. It is able to convert legal documents into a hypertext structure (Wilson, 1990).

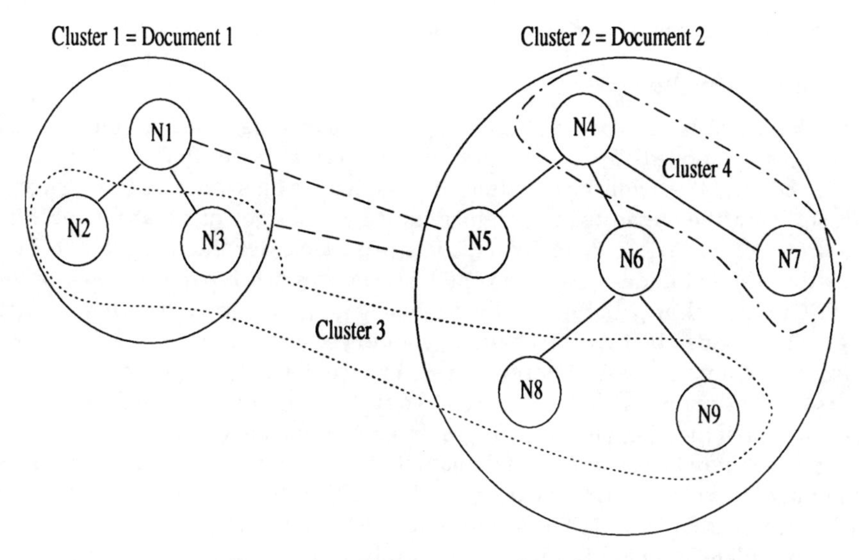

Figure 6.3: Clustering in a hypertext.

documents/clusters	nodes per document/cluster
1	N
K	M
N	1
x	y

Table 6.1: Number of documents/clusters in a hypertext depending on nodes per cluster.

clusters at a time. Figure 6.3 shows nine nodes (N1-N9) organized in two clusters (1 and 2) representing documents. One cluster (4) is part of a document (2) and one cluster (3) has nodes from two different clusters (1 and 2). The two documents are connected by a referential link (broken line), so are nodes 1 and 5. It could be possible as well to link a single node to a cluster.

The complexity of a hypertext does not only depend on the quantity of links between nodes and on the amount of nodes. Table 6.1 summarizes the possible number of documents depending on the amount of nodes per document in a network with N nodes. If the average number of nodes in a document is M there are approximately $K = N/M$ documents in the network. Furthermore, any number (x) of non-document clusters with any amount (y) of nodes per cluster can exist.

Looking at the potentials of hypertext for structuring information it becomes clear that implementing a retrieval system as a hypertext that does not store (large parts of) the actual information does not make much sense. Conventional information retrieval can be carried out better by the available IRS.

6.3.2 How to access hypertext for information retrieval

Querying vs. browsing

The netlike structure of hypertext asks for specific ways to traverse such an information network. As mentioned above, the user typically accesses node after node by following links that connect these information units. This activity is often referred to as *"browsing"*. When studying the literature about hypertext it becomes obvious that the meaning of this term is not well defined. However, some reasonable definitions exist. We will not discuss the best definition of browsing and related terms in this paper, but it seems to be necessary to cut a clear path through the confusing jungle of terms. In doing so, some of the benefits provided by hypertext systems should become clearer.

Users of a computer-based information system usually look for information for different reasons. The main distinction between two types of information needs is the answer to the question if the user has a specific *goal in mind* or if s/he wants more *global information*, e.g. to get a first orientation in a domain that is new for her/him. Having a specific goal in mind, it is obvious that the user should be able to access the needed information in a *direct* way without unnecessary effort or distraction. Therefore, some facility that lets the user formulate her/his information needs and that accesses and presents the retrieved information to the user has to be provided by the system software. Usually, this facility is a more or less specific query language combined with an access mechanism operating on the underlying data structure, e.g. a database. While conventional IRS provide this functionality, only a few of the commercially available hypertext systems allow querying so far. "Querying" and "travelling through hypertext" from node to node have both to be provided in order to satisfy the different types of information needs (see also Halasz, 1988, pp. 841 ff.).

Querying is one way to *search* for information and it implies the formulation of a search request and a *system driven* search for data that *matches* this request (see below for different retrieval models). However, searching for information first of all means that a user has a good idea (goal) of what s/he wants to find but the method how to access the information is not yet determined. Therefore, users who traverse a hypertext may be also searching for information, they just use a different access method than querying. On the other hand users might only be interested in a first orientation trying to cover a rather larger area of a domain without depth. We refer to this type of activity as *"scanning"*, where the goal is much less explicit compared to the activity of searching. When the goal becomes even more blurred or there is no explicit goal at all, a user may access different nodes of the hypertext in a somewhat unstructured manner looking for something interesting whatever it may be. We refer to this as *"wandering"*, a term McAleese uses for the activity of "purposeless and unstructured globetrotting" (McAleese, 1989, p. 11).

The activities "searching", "scanning", and "wandering" are strongly related to the user's intention and/or her/his momentary information need. According to the Collins COBUILD English Language Dictionary (Collins, 1987) browsing has quite the same meaning as wandering as we defined this term above (see Waterworth and Chignell, 1991, p. 36). We feel that browsing through a hypertext has a more general meaning as outlined in Kuhlen (1991, p. 126ff.). It can be differentiated between "directed" and "undirected" browsing.

Directed browsing (see also Hammwöhner, 1990) is characterized by a goal-driven search for information, e.g. in a hypertext, and is therefore another way besides querying of conducting a search process. Due to this relationship between searching and browsing Cove and Walsh (1988) call this activity "search browsing". One big difference to querying is the way the wanted information is accessed. After formulating a query, the system searches for data that match the query in some way and the user usually gets the number of "hits" and possibly a list of the retrieved sources. New information can only be found by reformulating the query and letting the system search again. When browsing through a hypertext, the search process itself is *user-driven,* because the user decides what path to follow and what information unit to access. This has great impact on the way how information is transferred to the user. When browsing through the information network the user may find information s/he hasn't had in mind when starting the search process. If the additional information is strongly related to the genuine information need this causes a more effective search process or a higher rate of retrieved information that is relevant. It may also happen that the user looses her/his former goal, because s/he finds more interesting information that is not strongly related to her/his formerly suggested information need. This is called "serendipity effect", something that obviously cannot happen when searching by querying.

The activities of "scanning" and "wandering" can then be summarized by the term "*undirected browsing*", also referred to as "general purpose browsing" (Cove and Walsh, 1988). Following associations between chunks of information is typical for undirected (associative) browsing (Kuhlen 1991, p. 130ff.). In our opinion "associative browsing" seems to be the natural way to traverse a hypertext, even if the user has a specific goal in mind, as long as the path to the goal is not predetermined.

Hypertexts cover a rather large part of or even the entire information and not only references to the actual information sources as conventional (non-fulltext) retrieval systems do. Therefore, users usually want to access this information by reading texts, by looking at graphics or video sequences, and by listening to audio data such as voices, music, etc. By doing this, the user tries to find out the extent of the stored information and travels through a specific region of the information network. This activity can be described as "exploring" (see also McAleese, 1989, p. 11).[4]

One term that is strongly connected to the activity of browsing is "*navigating*". While a user browses through a hypertext s/he has to keep track of where s/he is now (location), where s/he can go from here (direction), and what is the best way to reach a given destination (path).

Table 6.2 summarizes ways to access information depending on the user's intentions when s/he is looking for information. Obviously, querying is only well suited for searching that can be accomplished by directed browsing as well. So the user has to decide if s/he wants to formulate a search request using a query language or if s/he wants to browse through the hypertext, provided that both functions are available.

---

[4]Waterworth and Chignell (1991) use the term "exploring" in a more general meaning covering the activities of browsing and querying.

access method	directed browsing	undirected browsing	querying users intention
searching	X	–	X
scanning	X	X	O
wandering	–	X	–
exploring	X	O	–

X: well suited
O: possible
–: not possible

Table 6.2: Browsing/querying and users' intentions.

It should be stated again, that browsing and querying are by no means mutually exclusive functions of a retrieval system. The information needs of different users or of the same user at various times can be described as a continuum from searching to wandering. Therefore it is inappropriate to arbitrarily classify between user intentions and build systems that reflect this strict dichotomy (Waterworth and Chignell, 1991, p. 37).

This is even more relevant for hypertexts that store larger amounts of data, because for these "hyperbases" it is not efficient to start browsing without narrowing the search space considerably in advance. Especially for "Large Dynamic Hyperbases (LDHs)" that receive information from many sources on a continuous, dynamic basis, it is not sufficient to only provide browsing functionality. One way to cope with this problem is to build indices where each index entry can be seen as a "soft" or implicit link between nodes that contain this term, phrase, or descriptor. In comparison, hypertext links are "hard" or explicit and should be restricted in LDHs in order to reduce cost and effort when linking information units. Therefore, hypertext links may occur only in areas of a LDH where a rigid structure is mandatory and for associations that cannot be implemented via indices (Carmel et al., 1989). An approach how to build hypertext-based indices (hyperindices) using "index expressions" is outlined in Bruza (1990).

The way in which retrieval/querying functionality can be incorporated into hypertext systems is discussed in the next section.

Information retrieval models

The goal of information retrieval using a conventional IRS is to retrieve all relevant documents, while retrieving as few non-relevant sources as possible. Search requests often have to be incomplete or even vague, since (a) the user might not know her/his information need exactly, (b) s/he might not be able to express her/his information needs in a query language[5], and (c) there is an inherent problem to formalize search requests

[5]Most query languages apply boolean operators to connect keywords as representatives of the information the user is interested in to each other (see below). The formal structure of the language often causes problems to the novice and in some cases even to trained users. To avoid some of these problems a visual query

for (non-trivial) textual information. Information retrieval is distinguished from *data retrieval* which asks for exact and complete query specifications and where information units are retrieved using deductive inferences based on a deterministic representation. While in data retrieval exact matches to the search request can be found, information retrieval strives for partial or best matches (van Rijsbergen, 1979).

An information retrieval model is characterized by (a) the organizational structure in which the information units are represented and (b) a set of retrieval methods based on this organization. The standard retrieval model in most commercial (as opposed to experimental) retrieval systems is the *boolean model.* Queries are formulated as a set of descriptors or terms connected by boolean operators like "and", "or", "not". The document database is searched for the existence or non-existence of terms evaluating the boolean expression stated in the query. To improve the effectiveness of the search process, the database usually is physically organized as an inverted file system (Salton and McGill, 1983).

There are several disadvantages of the boolean model. Even experienced users find it difficult to formulate "good" queries resulting in an appropriate number of retrieved documents that are relevant for the user. The list of retrieved documents comes in random order, e.g. the order in which they entered the database. Due to the lack of mechanisms to express weighted terms, the user gets no information concerning the relevance of the retrieved documents to the formulated query.

Sophisticated retrieval techniques deal with similarity functions, weighted terms which represent documents and queries, organizing the document database in clusters and dynamic modification of queries and documents during the retrieval process (Salton and McGill, 1983). In the *vector space model* documents and queries are represented by vectors consisting of terms. The query and document terms may be weighted to express their relevance to the document or query. Based on a comparison of query and document vectors a degree of similarity is determined resulting in a ranked list of retrieved documents. The document database can be seen as a vector space whose dimension is equal to the total number of distinct terms.

A disadvantage of the vector space model is the lack of an explicit interpretation and underlying theory of the weights used in the vectors. This results in severe problems when trying to combine the model with more complex representations (e.g. structured documents; Fuhr, 1990).

To improve the effectiveness of the retrieval process the document database can be organized in *clusters.* A cluster is a set of similar documents. In a clustered database search can be constrained to those clusters evaluated as similar to a query. Clustering often results in hierarchical organizations which provide better recall and precision[6] relative to the cost of building and maintaining the cluster overhead.

Most conventional retrieval systems offer their users an interactive dialogue based on a decomposition of the retrieval process into phases (Salton and McGill, 1983). The user

---

language (GraphLog) was suggested by some researchers (Consens and Mendelzon, 1989).

[6] "Recall" and "precision" are the usual measures of retrieval effectiveness where *recall* describes the proportion of relevant documents which are retrieved and *precision* the proportion of retrieved documents being relevant (see van Rijsbergen, 1979, for a discussion).

is presented additional information in order to modify her/his query dynamically, e.g. during the retrieval process. Examples are a list of descriptors related to the originally expressed terms (e.g. taken from a controlled thesaurus), the "hit rate" stating the number of retrieved documents associated with the query, a list of document titles retrieved giving the user an impression of how the query is going to be interpreted by the automated search intermediary, etc.

Taking advantage of the *relevance feedback* users give during the retrieval process, some IRS are able to alter the weight of document descriptors in order to improve future retrieval effectiveness taking into account the interests of individuals or individual co-working groups.

One of the main constraints on sophisticated information retrieval is that the information in the document texts is expressed in natural language (Sparck Jones, 1983). Operating on document and query representatives instead on the objects themselves reveals the uncertainty inherent to textual information. In *probabilistic retrieval models* it is tried to deal with uncertainty by assigning probabilities to document-query pairs stating the probability that an arbitrarily chosen pair represented by vector terms will be considered as relevant by a user. The probabilities are received from heuristic functions, empirical studies or observations of the user's interest (Frisse and Cousins, 1989). In the Darmstaedter indexing approach the concept of a relevance description is introduced in order to get appropriate estimates of the probabilities based on a relative low number of observed queries (Fuhr, 1990).

In the following section several proposals for retrieval models dealing with hypertext structures are described.

Retrieval models for hypertext

Compared to traditional information retrieval there are two main differences when looking at hypertext. The notion of a document no longer describes an unstructured linear sequence of textual information but relatively small nodes organized as a network. Furthermore, as in conventional IRS, the retrieval process is not limited to document representatives (vector terms, titles or abstracts) but may take advantage of the fact that the documents' contents are represented in the network as well.

There are some requirements stated in the literature for integrating hypertext and information retrieval: access methods such as browsing in hypertext and querying in retrieval systems should both be offered to users in a highly intertwined manner (Fuhr, 1990). The transition between browsing and querying should be "smooth" allowing the user to change search modes as required by the respective status of the information seeking process (Waterworth and Chignell, 1992). Basic fulltext search features are provided by some well-known hypertext systems such as KMS (Akscyn et al., 1988), Intermedia (Yankelovich et al., 1988), HyperTIES (Shneiderman and Morariu, 1986), and Hypercard (Fraase, 1990), although more advanced approaches have been developed recently.

Frisse and Cousins (1989) describe an information retrieval model for hypertext. The entire information space consists of a *document space* and a separate *index space* that represents a structure of terms to describe queries and documents.

Contrary to traditional fulltext retrieval systems the information in the documents is divided into smaller units, represented in nodes and organized in a hypertext structure[7]. The nodes are related by various link types ranging from traditional (associative) reference links, hierarchical part-of links, through to semantical links, etc.

When starting a query, the user chooses a number of index terms leading to connected document nodes. While examining the document nodes s/he gives relevance feedback information to the system. This evidence is propagated in the index space by means of conditional probabilities existing between index terms, thus building a Bayesian network of belief values. In order to limit the complexity of value propagation the index is organized hierarchically. The result is a ranked order of nodes that might be interesting to the user. These nodes may be used as candidate starting points for a subsequent browsing process referred to by the authors as "local" search instead of "global" searching using terms from the index space in queries.

Although the learning system may not converge on the appropriate set of index space terms under conditions of uncertain information, the authors believe that even mildly misleading results will provide starting points for the use of subsequent access methods, e.g. browsing (Frisse and Cousins, 1989).

Croft and Turtle follow two different approaches to integrate information retrieval in hypertext. A deductive knowledge base representing the structure of a hypertext (adjacency of nodes, synonyms, reference links, etc.) is built using Prolog[8]. This approach may turn out as ineffective for large hypertexts. The reason for this lies in the uncertainty associated with natural language which affects the knowledge representation of the hypertext structure as well as the query (Croft and Turtle, 1989).

Therefore, maintaining the idea of information retrieval in hypertext seen as an inference process, an alternative approach by Croft and Turtle uses Bayesian networks. The main difference to Frisse and Cousins is that index terms (referred to as "concepts" by Croft and Turtle) are not separated from the documented space. This can cause problems due to the complexity of belief propagation in network structures[9] (Frisse and Cousins, 1989) and may prove more difficult to implement as a separate index space using available hypertext systems (Croft and Turtle, 1989).

The advantage of the integrated Bayesian network of Croft and Turtle is that reasoning can use relations (conditional probabilities) between index terms as well as relations between document nodes (adjacency, statistical neighbourhood, reference links, citations, part-of links). Pairs of any combination of concept and document nodes are supported.

Nevertheless, the approaches of Frisse and Cousins, and Croft and Turtle, have some properties in common: the use of Bayesian networks, the interpretation of information retrieval as a kind of *automated reasoning*, *user feedback* during the retrieval process to alter a query dynamically and the intention to use queries for identifying candidate

[7]Due to this decomposition contextual information is lost. Frisse (1988) describes how context can be represented by propagating numerical values in the network thus indicating that child nodes contain valuable information with regard to their parent node.

[8]In Lucarella (1990) multivalued logics are used as a framework to deal with uncertainty in a model for hypertext-based information retrieval

[9]Lucarella (1990) proposes the use of heuristics to constrain the search process, e.g. distance constraints.

starting points in the hypertext for subsequent browsing. Besides Bayesian networks, plausible inference (van Rijsbergen, 1986) and the use of Dempster-Shafer theory (Biswas et al., 1987) have been proposed in the literature to cope with uncertainty in information retrieval.

In Biennier et al. a model slightly different to those described above is introduced. The index nodes are connected by direct and reverse associative links in a "bidirectional neural network". Index terms that share the same or similar contents of document nodes are clustered and organized in a layered architecture. The layers represent different levels of specialization. The user can expand his/her query by specifying parameters such as the intended specialization level of the nodes' contents and the precision level for the index terms (Biennier et al., 1990)

This idea of querying to identify nodes/clusters to start browsing is also investigated by Crouch et al. (1990). They use a hierarchical clustered organization of document nodes based on similarity measures. An interactive browser provides the user with a graphical visualization of the cluster hierarchy. During the retrieval process the user is able to modify the query dynamically by adding or deleting index terms. Furthermore, s/he is able to examine the document nodes' contents to give better feedback to the retrieval system. In an experimental comparison with the SMART retrieval system[10] using the interactive browser resulted in better recall and slightly better precision. Certainly, a problem is that the user is responsible for the retrieval process and has to interpret a number of artificial numerical values to choose the appropriate clusters/document nodes (Crouch et al., 1990).

6.4 Conclusions

User's needs in information retrieval differ in a broad range that can be described as a continuum from very specific and structured search requests to a more or less aimless and unstructured wandering through the information base. To satisfy the requirements coming from these different needs an architecture for information retrieval systems has to be developed that is user-oriented and that allows effective and efficient information retrieval. Hybrid systems that combine the functionalities of IRS and hypertext systems seem to be a promising approach for this task. In this way conventional search by querying in IRS can be combined with the browsing capabilities of HTS. Queries are well suited to retrieve information when the user exactly knows what s/he is looking for and in order to find starting points for travelling through rather large information bases by following links. Browsing opens completely new ways to access information. Therefore, hypertext-based IRS promise far greater opportunities than conventional IRS. In addition to that, HTS usually offer sophisticated (graphical) user-interfaces that are easy to use.

Different approaches have been proposed how to integrate information retrieval and hypertext. Indices, eventually organized as hypertext themselves, can be used to directly access information units in a hypertext. Some (experimental) systems allow the user to give relevance feedback information to the system when examining retrieved

[10] For a description of SMART see Salton and McGill (1983), and Ellis (1990).

documents. This feedback is used to modify the query dynamically. Another approach based on relevance feedback is to apply automated reasoning using Bayesian networks and conditional probabilities between index terms as well as relations between nodes to retrieve information.

Still, some research and development has to be done in order to build effective and efficient hypertext-based IRS. Interesting projects that are underway are Envision and World Wide Web (WWW; see chapter 10, this volume). Envision is funded by the National Science Foundation and supported by the Virginia Polytechnic Institute and State University and the Association for Computing Machinery (ACM). In this three-year project it is aimed to develop an integrated software system combining object-oriented databases, hypertext, hypermedia, and information storage and retrieval capabilities. WWW is being developed at CERN in Geneva, Switzerland. Originally aimed at the High Energy Physics community it has spread to other areas. WWW merges techniques of information retrieval and hypertext to build a "global" information system.

References

1. Akscyn, R.M., McCracken, D.L. and Yoder, E.A., "KMS: a distributed hypermedia system for managing knowledge in organizations", *Communications of the ACM* **31**, 820–835, 1988.

2. Baeza-Yates, R., "String searching algorithms", in: Frakes, W. and Baeza-Yates, R., eds., *Information Retrieval: Algorithms and Data Structures*, Prentice Hall, Englewood Cliffs, NJ, 219–240, 1992.

3. Baeza-Yates, R. and Gonnet, G.H., "A new approach to text searching", *Communications of the ACM* **35**, No. 10, 74–82, 1992.

4. Bernstein, M., "An apprentice that discovers hypertext links", in: Rizk, A., Streitz, N. and André, J., eds., *Hypertext: Concepts, Systems and Applications*, Cambridge University Press, Cambridge, 212–223, 1990.

5. Biennier, F., Guivarch, M. and Pinon, J.-M., "Browsing in hyperdocuments with the assistance of a neural network", in: Rizk, A., Streitz, N. and André, J., eds., *Hypertext: Concepts, Systems and Applications*, Cambridge University Press, Cambridge, 288–297, 1990.

6. Biswas, G. et al., "Knowledge-assisted document retrieval: II. The retrieval process", *Journal of the American Society for Information Science* **38**, No. 2, 97–110, 1987.

7. Bloech, J. and Bogaschewsky, R., INFOR - Information-Retrieval-System, Working paper No. 1/1986, Institut für Betriebswirtschaftliche Produktions- und Investitionsforschung, Göttingen, 1986.

8. Bogaschewsky, R., "Hypertext-/Hypermedia-Systeme", *Informatik-Spektrum*, 15, 127–143, 1992.

9. Bruza, P.D., "Hyperidices: a novel aid for searching in hypermedia", in: Rizk, A., Streitz, N. and André, J., eds., *Hypertext: Concepts, Systems and Applications*, Cambridge University Press, 109–122, 1990.

10. Bush, V., "As we may think", *Atlantic Monthly* **176**, 1 (July), 101–108, 1945.

11. Carando, P., "SHADOW", *IEEE Expert*, 65–78, 1989.

12. Carmel, E., McHenry, W.K. and Cohen, Y., "Building large, dynamic hypertexts: how do we link intelligently?", *Journal of Management Information Sciences* **6**, No.2, 33–50, 1989.

13. Clitherow, P., Riecken, D. and Muller, M., "VISAR: a system for inference and navigation in hypertext", in: ACM (ed.): *Hypertext '89* – Proceedings Nov. 1989, Pittsburgh, PA, 293–304, 1989.

14. Collins, ed., *COBUILD English Dictionary*, Collins, London, 1987.

15. Conklin, J., "Hypertext: an introduction and survey", *IEEE Computing*, September, 17–41, 1987.

16. Consens, M.P. and Mendelzon, A.O., "Expressing structural hypertext queries in GraphLog", in ACM, ed., *Hypertext '89* – Proceedings Nov. 1989, Pittsburgh, PA, 269–292, 1989.

17. Cove, J.F. and Walsh, B.C., "Online text retrieval via browsing", *Information Processing and Management* **24**, 31–37, 1988.

18. Croft, W.B. and Turtle, H., "A retrieval model incorporating hypertext links", in ACM, ed., *Hypertext'89* – Proceedings Nov. 1989, Pittsburgh, PA, 213–224, 1989.

19. Crouch, D.B., Crouch, C.J. and Andreas, G., "The use of cluster hierarchies in hypertext information retrieval", in ACM, ed., *Hypertext'89* – Proceedings Nov. 1989, Pittsburgh, PA, 225-237, 1990.

20. Ellis, D., "A behavioural approach to information retrieval system design", *Journal of Documentation* **45**, 171–212, 1989.

21. Ellis, D., *New Horizons in Information Retrieval*, Library Assoc. Publ., London, 1990.

22. Engelbart, D.C., "A conceptual framework for the augmentation of man's intellect", *Vistas in Information Handling* 1, London, 1963.

23. Fraase, M., *Macintosh Hypermedia*, Vol. I, Reference Guide, Scott, Foresman & Co., 1990.

24. Frisse, M.E., "Searching for information in a hypertext medical handbook", *Communications of the ACM* **31**, 880–886, 1988.

25. Frisse, M.E. and Cousins, S.B., "Information retrieval from hypertext: update on the Dynamic Medical Handbook Project", in ACM, ed., *Hypertext'89* – Proceedings, Nov. 1989, Pittsburgh, PA, 199–212, 1989.

26. Fuhr, N., "Hypertext und Information Retrieval", in Gloor, P.A. and Streitz, N.A., eds., *Hypertext und Hypermedia*, Informatik-Fachberichte 249, Springer, Berlin, 101-111, 1990.

27. Gloor, P.A., *Hypermedia-Anwendungsentwicklung*, Teubner, Stuttgart, 1990.

28. Gonnet, G.H. and Baeza-Yates, R., *Handbook of Algorithms and Data Structures – In Pascal and C*, 2nd ed., Addison-Wesley, Wokingham, UK, 1991.

29. Halasz, F.G., "Reflections on Notecards: seven issues for the next generation of hypermedia systems", *Communications of the ACM* **31**, 836–852, 1988.

30. Hammwöhner, R., Ein Hypertext-Modell für das Information Retrieval, Thesis, Konstanz, 1990.

31. Hayes, P. and Pepper, J., "Towards an integrated maintenance advisor", in ACM, ed., *Hypertext '89*, Proceedings, Nov. 1989, Pittsburgh, PA, 119–128, 1989.

32. Hofmann, M., Benutzerunterstützung in Hypertextsystemen durch private Kontexte, Thesis, Braunschweig, 1991.

33. Kuhlen, R., *Hypertext*, Springer, Berlin, 1991.

34. Lucarella, D., "A model for hypertext-based information retrieval", in Rizk, A., Streitz, N. and André, J., eds., *Hypertext: Concepts, Systems and Applications*, Cambridge University Press, Cambridge, 81–94, 1990.

35. McAleese, R., "Navigation and browsing in hypertext", in: McAleese, R., ed., *Hypertext: Theory into Practice*, Ablex Publishing Corp., Norwood, NJ, 6–44, 1989.

36. Nelson, T.H., "Getting it out of our system", in: Schlechter, G., ed., *Information Retrieval: A Critical Review*, Washington, 191–210, 1967.

37. Nelson, T.H., "Managing immense storage", *Byte*, January, 225–238, 1988.

38. Newcomb, S.R., Kipp, N.A. and Newcomb, V.T., "The 'HyTime' hypermedia/time-based document structuring language", *Communications of the ACM* **34**, 67–83, 1991.

39. Nielsen, J., "Hypertext bibliography", *Hypermedia* **1**, 74–91, 1989.

40. Salton, G. and McGill, M.J., *Introduction to Modern Information Retrieval*, McGraw-Hill, New York, 1983.

41. Shneiderman, B. and Morariu, J., The Interactive Encyclopedia System (TIES), Department of Computer Science, University of North Carolina, 1986.

42. SIGLINK, Newsletter of the Special Interest Group on Hypertext (SIGLINK) of the Association of Computing Machinery, No. 1, ACM, New York, 1992.

43. Sparck Jones, K., "Intelligent retrieval", in Jones, K.P., ed., *Intelligent Information Retrieval, Informatics 7*, Aslib, London, 136–142, 1983.

44. Steinacker, I., *Dokumentationssysteme*, Springer, Berlin, 1975.

45. Stotts, P.D. and Furuta, R., "Hierarchy, composition, scripting languages, and translators for structured hypertext", in Rizk, A., Streitz, N. and André, J., eds., *Hypertext: Concepts, Systems and Applications*, Cambridge University Press, Cambridge, 180–193, 1990.

46. van Rijsbergen, C.J., *Information Retrieval*, 2nd ed., Butterworth, London, 1979.

47. van Rijsbergen, C.J., "A non-classical logic for information retrieval", *Computer Journal* **29**, 481–485, 1986.

48. Waterworth, J.A. and Chignell, M.H., "A model for information exploration", *Hypermedia* **3**, 35–58, 1991.

49. Wilson, E., "Integrated information retrieval for law in a hypertext environment", in *Proceedings of the SIGIR/ACM 11th International Conference on Research and Development in Information Retrieval*, ACM, New York, 1988.

50. Wilson, E., "Links and structures in hypertext databases for law", in Rizk, A., Streitz, N. and André, J., eds., *Hypertext: Concepts, Systems and Applications*, Cambridge University Press, Cambridge, 194–211, 1990.

51. Wu, S. and Manber, U., "Fast text searching allowing errors", *Communications of the ACM* **35**, 83–91, 1992.

52. Yankelovich, N. et al., "Intermedia: The concept and the construction of a seamless information environment", IEEE Computer, January, 81–96, 1988.

Chapter 7

Archie

Alan Emtage

Bunyip Information Systems
310 Ste-Catherine St. W, Suite 202
Montréal, Québec H2X 2A1 (Canada)
Email: bajan@bunyip.com

7.1 Introduction

The archie project provides a bonanza to the overused-cliche lover: "Necessity is the mother of invention", "Invent a better mousetrap and the world will beat a path to your door", "Be careful of what you ask for: you just might get it" are just a few that come to mind. In the computer vernacular, archie began life as a "hack" from a group of people who were trying to accomplish a goal with the least amount of work possible. It was originally conceived as a quick and dirty solution to a problem without much thought to design structure, prototyping, modularization or any of the other buzzwords so often heard in software engineering circles. Little did we know at the time that within 2 years of its release it would be the most widely (and often) used information system on the Internet today.

The pace of expansion of the Internet today is so rapid (doubling in size every 3 to 6 months) that the best we can hope to do is give you a snapshot of how we got here (difficult), where we are today (very difficult) and where we *think* it is going (akin to reading tea leaves).

7.2 What Archie Is

The Internet today provides access to many millions of bytes of information, much of which is available to its average user. However, it is often difficult to locate the data that one is looking for. The archie (Emtage and Deutsch, 1992) system is a widely known

A. Heck and F. Murtagh (eds.), Intelligent Information Retrieval: The Case of Astronomy and Related Space Sciences, 103–111.

tool for locating programs, documents and datasets. The current archie system (version 3) provides a general purpose method of collecting arbitrary information from widely distributed sources which is then incorporating it into a configurable set of databases. Users may then query the system to search for the desired data, through a number of archie client programs many of which are freely available on the Internet. In conjunction with other Internet information systems, archie also provides a *white pages* service with the ability to locate information about people and institutions on the network as well as a *yellow pages* database containing information about specific available services.

7.3 History

The Internet was originally intended to facilitate communications among researchers and initially provided such basic services as electronic mail, file transfer and remote login. FTP, the File Transfer Protocol (Postel and Reynolds, 1985) was developed very early on in the network's history and was designed for the reliable transfer of files across the network, complementing the telnet protocol which provided remote interactive capabilities.

Over one quarter of the traffic on the National Science Foundation network (NSFnet) backbone in the United States currently is ftp. The protocol was originally designed to allow file transfers for those people who had accounts on multiple computers on the network. So for example, if you were a researcher travelling in California but your home base was in New York, you could use ftp to log on and retrieve those files assuming you had access to an account on a machine in California. However, it is often the case that you would like to make certain information generally available to the average network user.

The usual ftp session begins with a standard login procedure with the use of a *login name* to identify yourself to the remote machine followed by a password to confirm your authenticity as that user. The use of *anonymous ftp* was developed as a convention whereby using the special login name *anonymous* would allow you access to selected files on those particular sites which had the feature enabled. No password would be requested. Thus anyone who knew the procedure as well as those machines which employed it, could access the available public files. The sites using this convention became to be known as *archive sites* on the network. Thus if you had, for example, published a paper which you wanted to make available to any of your colleagues with network connectivity, you could ask the administrators of your favourite archive site to add your file to their collection.

By 1989 there were about 600 of these archive sites worldwide. They ranged in size from a few kilobytes to several hundred megabytes of available material. Some served as general purpose repositories for much of the public domain and freely available programs, documents and datasets on the network. Others were more specialized (such as those at NASA, which contained such things as the shuttle launch schedules, images from previous space missions and the latest status reports on many of the ongoing NASA projects). Many sites were professionally run and maintained while others were the pet projects of individual site administrators with a little extra disk space to spare.

There was however a problem with this ad hoc system: how would one locate the information that one needed? The dissemination of the availability of particular pieces of information was often conducted by word of mouth and if was often very difficult to locate the particular bit of data that you were looking for. Entire mailing lists and Usenet newsgroups were created for people asking for the software or documents that they were looking for and knew resided on an archive site somewhere and even with this, it was often the case that one would not get an answer.

The archie project was started as an attempt to provide local system administrators of the School of Computer Science at McGill University with the ability to quickly search through the contents of the more popular archive sites when trying to locate a particular piece of public domain software on the network. A simple set of automated scripts would periodically log onto a predetermined list of archive sites and request a listing of all the available files, along with their location on the site, their last date of modification, size and other information. These listings were then placed in a common location on one of the machines in the department. Individual files could then be located by any one of a number of string matching programs. Such a system of course assumed that you knew, or could at least take a guess at, the name of the file that you were looking for. The system was only minimally advertized inside the department and was mostly used by system administrators in the course of their work. The turning point in the project was when the School's System Manager answered a request from an Internet user on the Usenet system for a particular program. His explanation of how the information was obtained revealed the existence of archie to a larger community and over the course of the next several days, a number of people often asked the staff to perform additional searches. When it had reached the point that the work of the group was being disrupted, it was decided that a simple network interface would be provided to allow any user on the Internet to log into a "dummy" account on the machine containing the listings and perform the searches themselves. This was the original archie[1].

Once it became clear at the end of 1990 that the service was gaining in popularity (at an exponential rate), the interface was rewritten in a more powerful programming language and with significantly more features than the original version. In addition to the original telnet method of using the system, an electronic mail (email) interface was added, allowing users a second method of accessing the database. In addition a prototype Software Description database was added containing short descriptions of over 3,500 programs, documents and datasets then available on the Internet. Users could perform a keyword search on the database when trying to locate the name of a particular file which might contain the information that they desired. This database was manually constructed and maintained from a number of indexes widely available on the network.

In November 1990 there were approximately 50 logins per day at the original archie site. By March 1991, there were over 500 logins per day as well as another 40/day via the email interface. By June of 1991, 50% of all network packets on the link coming to Montreal from the United States were destined for the single archie machine (a Sun Microsystems Sparc 1+ with 16Mb of RAM). This link, which also serves all of Eastern

[1]The name *archie* derives from the word *archive*

Canada was effectively saturated. This was enough to bring us to the attention of the network authorities who were not at all pleased. The code for the system was distributed to other archie sites starting with the regional and national networks, FUnet in Finland, SURAnet and ANS in the USA and AARnet in Australia which greatly helped to spread the load. By October of 1991 archie was being described as one of the basic Internet services. Although statistics are currently difficult to gather, it is estimated that as of June 1992, archie servers worldwide, from Japan, Australia, New Zealand, North America and Europe service over 50,000 queries per day with the system tracking about 1600 archive sites worldwide. These archives contain an estimated 2.5 million files with a combined data storage of in excess of 170 Gigabytes (170,000 million bytes). In January 1992, work on archie was taken over by Bunyip Information Systems, Inc., a company created to work on archie and other Internet information technologies.

7.4 Information Systems on Very Large Networks

In order to understand what archie does and how it fits into the wider picture of information systems on the Internet, it is necessary to describe the larger conceptual framework in which it exists.

Over the past several years a significant amount of research has been done on the problems associated with trying to build a ubiquitous, easily used information infrastructure (or *infostructure*) on computer networks. Professor Mike Schwartz at the University of Colorado, Boulder, has identified three levels which the user need navigate in order to locate the particular piece of information desired (Schwartz, 1991). These are:

- Class Discovery

  This refers to the user having to locate a particular group of resources in the network environment. For example, all archive sites on the Internet may be described as belonging to the same "class" of objects. At the start, users may not even know that a particular class of resources exist.

- Instance Location

  Once the class of objects desired has been located, a particular instance of that class needs to be discovered. Again, with archive sites, one must locate a *particular* archive site or sites with the desired information to continue with the search.

- Instance Access

  Once the instance of the object has been located there are an associated set of problems with accessing that "instance" (the particular archive site(s) and the information that you desire). Archive sites may seem to be a simple example of this since almost by definition one knows the access method. However those supporting archie have often communicated with users who have discovered archie and have been able to use the system, but lack the knowledge necessary to obtain the data even when they have located what they were looking for.

In the past two years the archie system has mainly addressed the problem of Instance Location, in that it allows one to find those files on known archive sites on the Internet. The underlying assumption is that the Class Discovery has been implicitly done by the archie system through notification to the archie maintainers by the administrators of the archive sites themselves. Archie does not attempt to locate new archive sites on its own and will only track those sites which have specifically approved of the operation. In addition there is the "initial contact" problem: one must have a point of contact to tie into any of the information systems currently available. While this is usually not a problem for those people who have been working in the Internet environment for a little while, it can be very difficult for the novice user, and this problem will only get worse as the size and complexity of the network increases.

7.5 Models of Information Gathering

The archie system operates, as described above, by contacting its set of archive sites on a regular basis (*polling*) and obtaining the desired information. This information is brought back to the server and incorporated into a single database that is then server to the entire community. This *centralized* model can be compared with the WAIS system (Kahle, 1989) which works on a much more distributed model whereby the information remains distributed among many servers: the user chooses a set of servers (from a list provided from a centralized source) which might contain the desired data. The user (through the client) then requests each server in turn to search for the desired data. It turns out that these models, rather than compete, complement one another depending on the type of data served. If searching the data is "expensive" (in terms of system resources such as CPU usage and memory), but gathering it is not resource intensive, then the archie model works well. On the other hand, if the search methods do not consume many resources but collecting the data is difficult, then the distributed model is the most appropriate avenue.

The integration of archie with other Internet information systems is an area under active investigation, as will be described later.

7.6 How Archie Works

Conceptually, the archie system is quite simple, consisting of a number of basic components which work together to collected, collate and serve the tracked information. The interfaces between the components have been defined so that administrators who wish to have the system track and maintain information not in the core distribution may do so without having to modify the basic system itself.

The archie system maintains two distinct classes of data.

- **Host Databases.** The so-called *host databases* which contain information about Internet hosts which make information available (the *Data Hosts*). Each record for such a host includes such things as

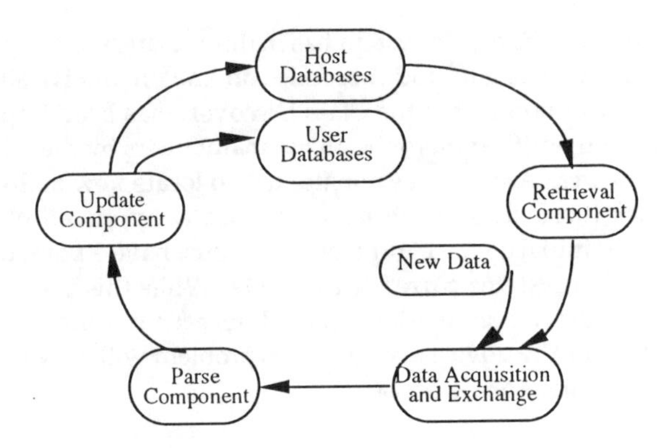

Figure 7.1: The update cycle.

- the name of the Data Host
- the operating system. (This determines how some functions are to be performed)
- the method of access for the resource (such as ftp, WAIS, telnet)
- commands to be performed to actually retrieve the data from that host
- the timezone in which the host resides. (This is used to schedule retrieval operations at off-peak times for the host and allow Internet users to convert times given by that host into their own local time)

Other administrative information such as times of last retrieval and update of the data that the host provides is also maintained.

These host databases can be considered the core of the system since they contain all the information necessary to drive the other components.

- **User Databases.** The *user databases* contain the actual information to which the user queries are directed. Current examples include the popular anonymous ftp listings and Software Description databases. The internal structure of user databases are not defined by the archie system itself which need only be provided with programs to perform the normal maintenance routines such as insertion and deletion of the data. As will be explained later, the archie system also does not define the mechanisms to search and access the data. This approach allows any number of arbitrary query methods to operate on the user databases.

7.7 The Architecture

There are four phases in building any of the user databases in the archie system, and for each there is a related archie system component. The process of collecting, collating and integrating the information into the user databases is known as the Update Cycle which is illustrated in Figure 7.1. In addition to retrieving the information from the Data Hosts themselves, the data exchanges are also performed between other Internet hosts running the archie system. This allows each archie to be *directly* responsible for only a fraction of the Data Hosts on the Internet while serving data collected by other archie systems.

- The Retrieval Component queries the Host Databases about which Data Hosts are to be updated and fetches the needed information.
- The Data Acquisition component is concerned with obtaining any new data (directly from the Data Hosts if the archie host in question is responsible for that site) or for obtaining the information from another archie host if it is not.
- The Exchange Component, using a basic flooding algorithm similar to the one used in the Internet Usenet news system, is used to maintain weak consistency between an arbitrary number of archie servers. Virtual networks of archie systems can be created and any individual server may belong to zero or more of these networks working together to exchange data.
- The Parse Component is responsible for parsing the incoming data into a form suitable for the update of the databases. This often means removing extraneous information (filtering) as well as performing data conversions, collation and compression.
- Finally, the Update Component obtains the parsed data and integrates it into the archie databases. This includes updating the Host databases as well as entering the new data obtained from the Data Hosts (or from the other archie hosts).

Figure 7.2 illustrates the basic components of the archie system and how they are related.

Each phase has a controlling routine called a Component Manager which is responsible for coordinating the activity in its component. Administrative information is attached to and associated with each piece of data passing through the Update Cycle. With reference to this information, each component can determine what particular routines need to be invoked to perform the necessary data transformations. This means that the upper levels of the system are unaware of the structure of the underlying data since those data transformations are performed by individual routines specifically designed for those functions.

7.8 Serving the Data to the User

Although the core archie system is concerned with collecting and incorporating the data from Internet hosts, the ultimate aim of course is to provide this data to users. The archie system itself has no predefined protocols for this function, rather it utilizes other Internet information systems to serve its information to the Internet community. For

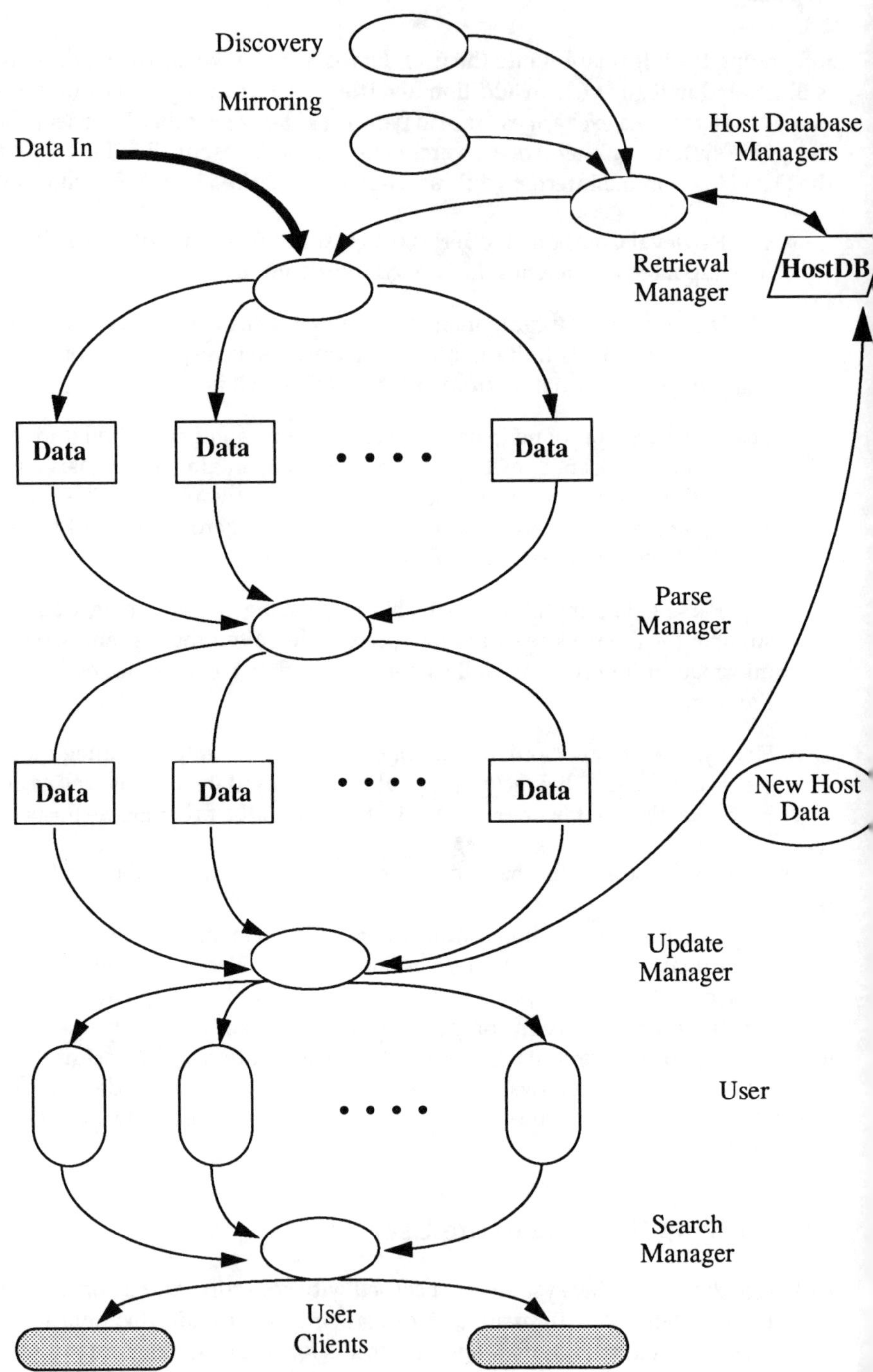

Figure 7.2: The archie architecture.

example, the Prospero system designed and developed by Clifford Neuman (1992), now of the Information Sciences Institute of the University of Southern California, has been instrumental in making the archie system what it is today.

Prospero provides an Internet-wide distributed file system whereby selected files on an Internet host can be made visible to the rest of the Prospero system. Since the original archie anonymous ftp databases consisted of file listings, the interface between the systems worked very well. All of the freely available archie clients work by contacting a Prospero server running on an archie host. The Prospero server has routines to search the anonymous ftp listings database and return this information to the client. The archie system is responsible for collecting the data and the Prospero system handles the searching and client/server interactions.

The Wide Area Information Servers system now also provides access to many of the databases that archie maintains and it is expected that other systems, such as Gopher (McCahill, 1992, and chapter 9 in this volume) and the World Wide Web (WWW) hypertext project at CERN (Berners-Lee et al., 1992, and chapter 10 in this volume), will also be directly integrated into archie in the not too distant future.

7.9 Getting More Information

Manual pages and other information are available from many Internet archive sites and often from your local administrator or by sending electronic mail to `info@bunyip.com`.

References

1. Berners-Lee, T., Cailliau, R., Groff, J.-F. and Pollermann, B., "World-Wide Web: the information universe", *Electronic Networking*, Meckler, Spring 1992.

2. Clifford Neuman, B., The Virtual System Model: A scalable Approach to Organizing Large Systems, PhD thesis, University of Washington, Department of Computer Science and Engineering, 1992.

3. Emtage, A. and Deutsch, P., "archie – an electronic directory service for the Internet", *Proceedings of the Usenix Technical Conference*, 93–110, January 1992.

4. Kahle, B., "WAIS – The Wide Area Information Server concepts", Technical Report TMC-202, Thinking Machines Corp. November 1989.

5. McCahill, M., "The internet Gopher", *Proceedings of the 23rd Internet Engineering Task Force*, San Diego, Ca., 1992.

6. Postel, J. and Reynolds, J., "RFC 959: File Transfer Protocol", ISI, October 1985.

7. Schwartz, M., "Resource discovery in the global internet", Report CU-CS-555-91, University of Colorado at Boulder, November 1991.

Chapter 8

WAIS

Jim Fullton
Clearinghouse for Networked Information Discovery and Retrieval
PO Box 12889, Research Triangle Park, NC 27709 (USA)
Email: jim.fullton@cnidr.org

8.1 Introduction

The promise of easy access to large amounts of archival data has long been of interest to astronomers. The continued development of high-speed international networks, relatively inexpensive mass-storage units, and fast, low cost computers have made such access possible. In conjunction with the evolution of the communications infrastructure required to meet this goal, a remarkable set of mostly experimental software tools have been developed. One of these tools is the Wide Area Information Server, commonly known as WAIS.

8.2 What is WAIS?

WAIS is a search and retrieval system loosely based on the ANSI protocol Z39.50-1988. Originally designed by Thinking Machines, Inc. as a free-text searching system, WAIS has been extended by many developers to provide support for non-textual objects such as astronomical images. The design of the WAIS system removes the need for direct correspondence between searchable data and retrievable objects. Thus, graphical objects such as images can be located based on the relational information found in the image headers, textual information contained in observers' notes and journal articles, or other information sources.

WAIS can be thought of as a "fine-grained" client/server search mechanism in which the client is tightly integrated with the platform upon which it operates. The server provides a search mechanism carefully tuned for the target information. The result is a

A. Heck and F. Murtagh (eds.), Intelligent Information Retrieval: The Case of Astronomy and Related Space Sciences, 113–117.

searching system that can quickly locate an object within a vast collection, provided the search engine is given enough clues to narrow the choices sufficiently.

The main decision-maker in the WAIS system is the user; only the user can determine the relevance between a piece of information and the query used to find it. Because of this, WAIS clients usually present a ranked list of objects to the user based on the given search criteria. This allows the user to select the most useful object or to mark an object as interesting and refer it back to the server for search refinement.

8.3 Current Information Access Mechanisms

Astronomical research generates a rich set of information. In addition to data that can be retrieved from various fixed storage media, some of this information is available only in paper format. While each of these formats is useful, both present challenges to researchers who choose to focus on archival information.

Paper archives most commonly consist of journal volumes. While the printed journal is the preferred format for the exchange of scholarly information among astronomers, the astronomical societies must arrange to print and distribute thousands of identical copies on a regular basis. Recipients of this printed matter then invest large sums of money for buildings and shelf space to store these volumes in a central location. As this central location is frequently inconvenient to the individuals who actually use these materials, they arrange to receive individual copies and store them in satellite locations, such as their own offices. While some find it important to have a paper journal "in-hand" for browsing, many readers would find great value in having the same information available in electronically searchable form.

However, there are equally difficult problems associated with current non-networked electronic mass information distribution methods. Virtually all astronomical images and databases are available on magnetic tape or optical disk. While inexpensive, magnetic tape is awkward to store and ship and quite slow for anything other than pre-planned sequential access operations. Tape is available in many formats, and astronomers have seemingly availed themselves of every possible combination of data types and tape formats. Researchers may find themselves in the position of having a needed piece of information locally available on tape, but without the device required to read it.

Additionally, information shipped on physical media is static; it can only be updated through conscious effort on the part of the user. Any reasonable upgrade mechanism must rely on redistribution of information on the same medium, or intervention by the user to merge new information into the old information base. This is a serious problem with optical disks which must be re-mastered for each change in the information base. While this is acceptable for fixed archival information, it becomes quite expensive to distribute dynamic information with this technology. Optical disks also require moderately expensive disk players for each workstation.

Keeping information at an electronically accessible central site seems to remedy these problems. In the past, this has been precluded by poor network connectivity and the cost of mass-storage units. Improved connectivity of the international network combined with the rapidly decreasing cost of data storage and the proliferation of desktop computers provides a solution to these access problems.

8.4 The Use of WAIS in Astronomy

WAIS has been successfully used to provide access to astronomical literature through the NASA Study of Electronic Literature in Astronomical Research Project (STELAR). STELAR provides electronic access to approximately 20 years of abstracted astronomical journal articles through full indexing of each word in the abstract. A query consisting of a list of words will be matched by any abstract containing any of the words. This mechanism is superior to basic keyword searching because the process of condensing an abstract into a small number of keywords results in the loss of important contextual information. Because the author-defined keywords are still available and included in the searchable text, fast text searching provides results that are at least as useful as those obtained using older methods.

The collection of STELAR abstracts is stored on a central mass-storage unit at Goddard Space Flight Center in Greenbelt Maryland, USA. Researchers may access journal abstracts through a variety of clients available for many popular computers. The user enters a query which is presented to the server by the client. The server removes common "stop" words such as "with" and "it" and then searches for each remaining word in the query, ranking the matching abstracts based on the frequency and position of each word within the text. This ranking is constrained to fall between 0 and 1000, and the list is returned to the client along with information about the formats in which each abstract may be retrieved.

The client presents this list of "hits" to the user, who may select any item of interest. The requested abstract is then returned in the format selected by the user. The system has been expanded on an experimental basis to allow the user to retrieve the actual bitmapped pages of selected journal articles. This is a powerful information selection resource; while textual material is searched, non-textual objects may be returned based on user-selected format criteria.

The same mechanism has been extended to primarily non-textual data. As an experiment, NASA has made selected elements of the International Ultraviolet Explorer (IUE) satellite archive available through WAIS. IUE spectra are stored on a 1.7 terabyte optical storage unit at Goddard. The unit is controlled by a Digital Equipment Corporation VAX running the VMS operating system. A VMS WAIS server accepts queries from clients and presents them to a database module which returns a list of appropriate spectra. The user selects a format and retrieves the data across the network.

Each of these archives may be accessed by free, public-domain WAIS clients, without the need for site licenses or proprietary software. The WAIS system provides a programming environment that allows developers to create clients customized to particular applications and servers tailored for particular archival problems.

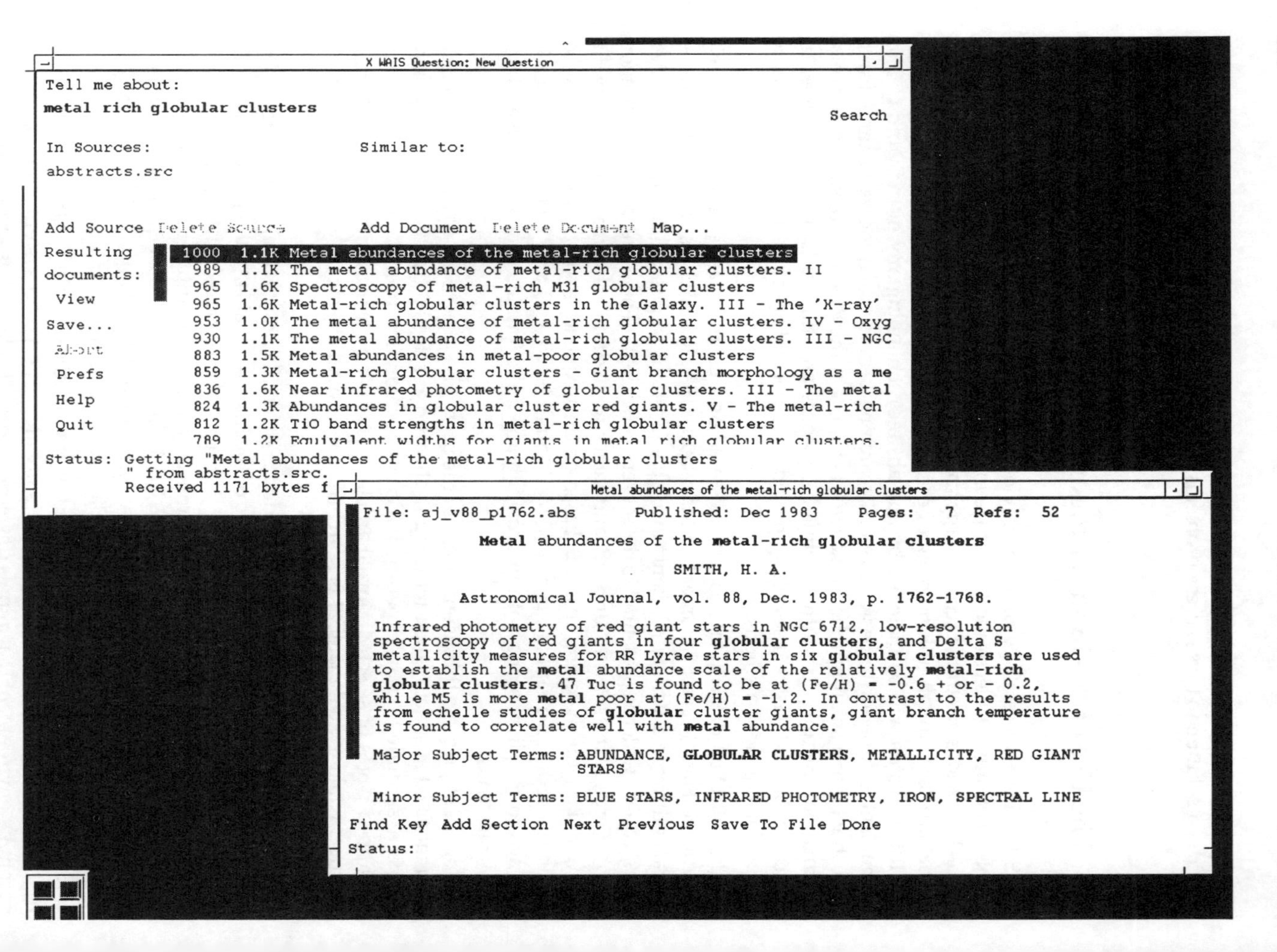

Figure 8.1: A WAIS session.

Public domain WAIS server and client software is available from a wide variety of sources by anonymous ftp:

Basic distribution – includes server, text search engine, text indexer, X client, and character mode client:
`wais-8-b5.1.tar.Z` from `think.com` and `sunsite.unc.edu`
Note that the version numbers are subject to change.

Additional Clients:

Macintosh – `WAISStation` from `hypatia.gsfc.nasa.gov`
Motif – `mxqwais` from `ftp.eos.ncsu.edu`
OpenLook – `openlook.tar` from sunsite.unc.edu
Microsoft Windows – `winwais.zip` from `sunsite.unc.edu`
MS-DOS – `pcwais.zip` from `sunsite.unc.edu`

Information about the STELAR project (including access instructions) can be requested by email from `stelar-info@hypatia.gsfc.nasa.gov`

General information about networked information tools is available from:

The Clearinghouse for Networked Information Discovery and Retrieval
MCNC Center for Communications
P.O. Box 12889, Research Triangle Park, NC USA 27709
Email: cnidr@cnidr.org

Table 8.1: Accessing WAIS

8.5 The Future of WAIS in Astronomy

WAIS has the potential to become a low cost, highly interoperable information distribution system for the entire astronomical community. The WAIS system itself is completely independent of search engine implementations, and the servers and clients can be developed and modified independently of each other and of the information managed by the servers. Experimental clients already exist for commonly used image reduction applications such as IRAF, and as more archives provide WAIS access, new clients will be developed. Furthermore, WAIS is not a "single-server" system. WAIS servers can easily co-exist with other data distribution systems.

Experiments with WAIS have shown it to be a low cost, robust and easily modified information system well-suited to the archival problems of astronomical researchers. The "open systems" nature of WAIS makes it a useful choice as either the primary or secondary information dissemination mechanism for data archive sites.

Chapter 9

The Internet Gopher

Farhad Anklesaria and Mark McCahill

Microcomputer Center, 132 Shepherd Labs, 100 Union Street SE
University of Minnesota, Minneapolis MN 55455 (USA)
Email: fxa@boombox.micro.umn.edu, mpm@boombox.micro.umn.edu

9.1 Internet Duct Tape?

Internet Gopher is a distributed server document search and retrieval system. Gopher combines the features of both electronic bulletin board services (a hierarchical organization of items) and of full-text searchable databases (searches based on the content of documents where all words in the document are considered keywords). The Internet Gopher system is based on a client/server architecture so users on a heterogeneous mix of desktop systems can browse, search, and retrieve documents residing on multiple distributed server machines anywhere on the Internet.

While documents (and services) reside on many servers, Gopher client software presents users with a hierarchy of items and directories much like a file system. In fact, the Gopher protocol and the usual Gopher client interface is designed to resemble a hierarchical file system since a hierarchy is a good model for locating documents and services: application developers find it easy to write and debug clients and servers; users need no training in using client software. The Gopher protocol is also extremely flexible, so services such as ftp, WAIS, archie, finger, whois, or NNTP can all be accessed from Gopher (using the familiar Gopher interface). Terminal session based resources typically don't follow a client/server protocol, but terminal-based information services can be cataloged, located and appropriate applications (such as telnet) launched for users from a gopher client. Because of this natural unifying role, Internet Gopher can be considered as an "Internet Duct-Tape". The duct tape seems to be deployed liberally: over the year 1992, Gopher grew from the 200th most popular protocol to number 16 on the list, based on packet counts on the NSFnet: this jump was made by a 4400 fold increase in its use across the NSFnet.

A. Heck and F. Murtagh (eds.), Intelligent Information Retrieval: The Case of Astronomy and Related Space Sciences, 119–125.

Gopher client and server software is available via anonymous ftp or via Gopher. By anonymous ftp from `boombox.micro.umn.edu`; look in `/pub/gopher`; if you are using Gopher, go to the University of Minnesota's main Gopher server and look in the Information About Gopher directory. Client software is available for Macintosh, PC-DOS, Unix, NeXT, X, VAX/VMS, VM/CMS. Server software is available for the Macintosh, PC-DOS, Unix, NeXT, X, VAX/VMS, VM/CMS, and MVS systems.

If you have an account on a timesharing Unix system, it may well have a Gopher client installed; try typing `gopher`. If you aren't able to get at a Gopher client in any of these ways, you can try telnetting to one of the following publicly accessible client services.

Hostname	IP#	Login	Area
consultant.micro.umn.edu	134.84.132.4	gopher	North America
gopher.uiuc.edu	128.174.33.160	gopher	North America
panda.uiowa.edu	128.255.40.201	panda	North America
gopher.sunet.se	192.36.125.2	gopher	Europe
info.anu.edu.au	150.203.84.20	info	Australia
gopher.chalmers.se	129.16.221.40	gopher	Sweden
tolten.puc.cl	146.155.1.16	gopher	South America
ecnet.ec	157.100.45.2	gopher	Ecuador

It is recommended that you run the client software instead of logging into the public login sites. A client uses the custom features of the local machine (mouse, scroll bars, etc.) and gives faster response.

Table 9.1: Getting your own Gopher.

9.2 Surfing Gopherspace

Internet duct tape sounds rather quaint and abstract, so a concrete description of surfing the Internet with Gopher should make things clearer. It would be ideal if you can obtain and run a Gopher client yourself while following this text. See Table 9.1, "Getting Your Own Gopher". The following describes a session using the TurboGopher client software for Macintosh; the user-interface details are Macintosh, but they map in a straightforward fashion to client interfaces on other machines.

Upon launching the client software, a network connection is made to some "home" Gopher server. This could be a local server at your University or corporation, or it may well be one of the University of Minnesota's gopher servers; the beauty of the Internet is that it really doesn't matter. Most clients are preconfigured to know about how to reach some "home" server. The client retrieves a listing of the top level directory from

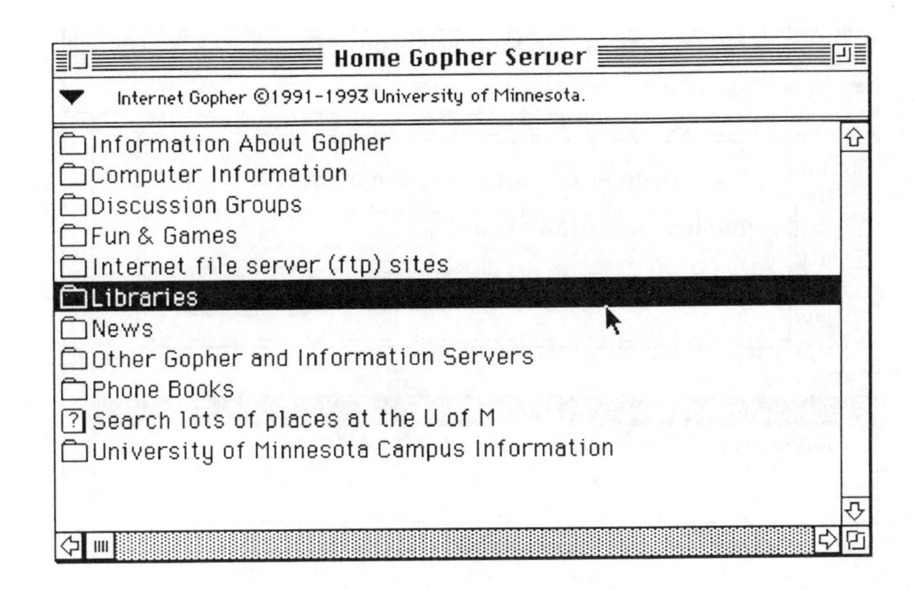

Figure 9.1: Top level directory

the server. This may appear something like Figure 9.1. On a Macintosh client, directory items are represented by different icons (e.g. the icon of a document or of a folder). On other platforms the directory lists may appear as a scrolling list of menu items; short text tags such as a trailing "/" or <DIR> may be used to denote a directory. In any case, to view the contents of a file, the user can just double-click on the file (or select it and then press Enter or Return on other platforms). Viewing the contents of a directory are equally straightforward.

It is not evident however that what the user sees is a "virtual" directory; items in the listing may be on many different machines anywhere on the Internet! As the user clicks from one directory to another and browses one file after another, he goes from one machine to another, exploring a huge networked filesystem, without knowing anything about commands, hostnames, or other network incantations. The client and server transparently manage the details of where the items are located.

There are other item types besides files and directories in Gopher directory listings. One of them (a search engine) is shown by the question-mark icon in Figure 9.1. By selecting a search engine item the user can submit to the server a search request for documents containing certain words (Figure 9.2). Most such search servers maintain full-text indexes of some subset of gopherspace. The "full-text" means that every word in every document is a keyword. The "index" means that the server can rapidly (a few

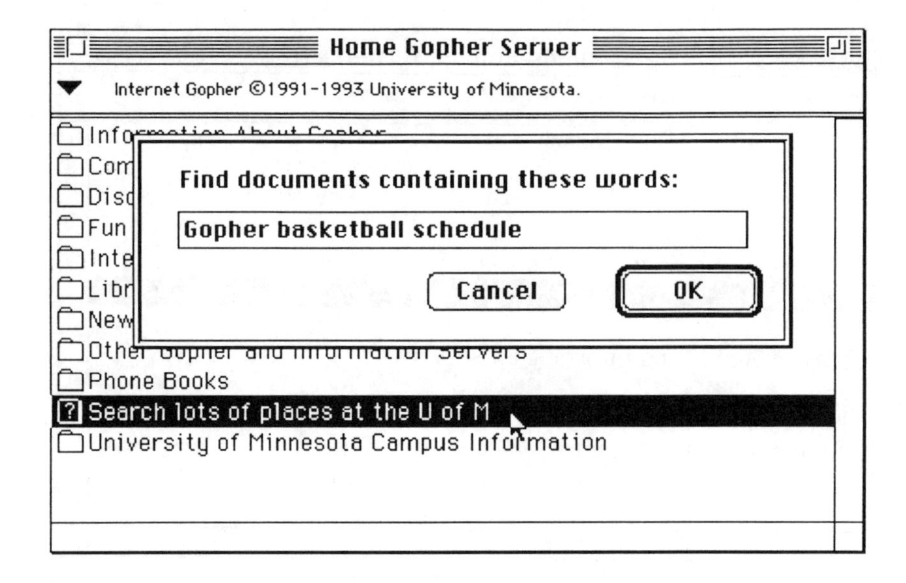

Figure 9.2: Search request.

seconds or less) return the user a list of all documents that match the search criteria. What the user sees as a result of the search is a list (or directory) with these documents in it; the server has simply sent the list to the client and the client has displayed it to the user in a directory that "resulted" from executing the search.

You may search the titles of all documents in networked gopherspace by using a search engine named "veronica". Veronica is an acronym for Very Easy Rodent-Oriented Net-wide Index to Computerized Archives. Rather than basing its search on the contents of documents, the Veronica search index is built from the title of items in gopher directory listing. This makes Veronica a good way of locating items if you have some idea of their name in Gopherspace.

Gopher can access services built on other client/server protocols such as WAIS or archie servers (see chapters in this volume) transparently: to the Gopher user these services look like gopher search engines. Anonymous ftp servers look like directories to the Gopher user. In each case, Gopher insulates the user from the oddities of WAIS, archie, and ftp servers and clients.

While the lowest common denominator is text, users can also view graphics files, hear sound files or retrieve software for their desktop machines in a variety of formats (binary, Macintosh BinHex, uuencode).

There are a number of services on the Internet that cannot be mapped onto the Gopher paradigm because they are so ad-hoc or because they insist on managing their own arbitrary interfaces. The best example of such services are most of the library catalogs accessible across the Internet via telnet sessions (based on vt100 or tn3270 terminal emulation). Remembering the domain names, ports and terminal emulations of all such services is a task that most users would rather avoid. While Gopher cannot hide the terminal-based interfaces presented by such services, it can catalog lists of such services and when the user selects one, Gopher can launch a telnet session to the host machine. Is the moniker "internet duct-tape" beginning to sound accurate?

9.3 Gopher Guts

The Internet Gopher protocol is remarkably simple – almost spartan. A client connects to a server and sends the server a selector (a line of text, which may be blank) via a TCP stream. The server responds with a block of text and immediately closes the connection. No state is retained by the server between transactions with a client, and typical transactions last under a second. The simple nature of the protocol stems from the need to implement servers and clients quickly and efficiently for slower, smaller desktop computers (1 MB Macs and DOS machines).

Below is a simple example of a client/server interaction. The client software knows about one "home" server; that server can then direct the client to all the other servers in gopherspace. The only configuration information the client software retains is this servername and port number (in this example that machine is `rawBits.micro.umn.edu` and the port 70). In the example below the # denotes the TAB character.

```
Client:              (Opens connection to rawBits.micro.umn.edu at port 70)
Server:              (Accepts connection but says nothing)
Client: <CR><LF>     (Sends an empty line: Meaning "list what you have")
Server:              (Sends a series of lines, each ending with CR LF)
0About Gopher#Stuff:About internet Gopher#rawBits.micro.umn.edu#70
1Around the University of Minnesota#Z,5692,AUM#underdog.micro.umn.edu#70
1Microcomputer News & Prices#Prices/#pserver.bookstore.umn.edu#70
1Courses, Schedules, Calendars#/Courses#events.ais.umn.edu#120
1Student-Staff Directories##uinfo.ais.umn.edu#70
1Departmental Publications#Stuff:DP:#rawBits.micro.umn.edu#70
                     (.....etc.....)
..                   (A period on a line by itself)
                     (Server closes connection)
```

The first character on each line tells what type item the line describes (for example: document, directory, CSO phone book server, or error are characters '0', '1', '2', or '3'; there are a handful more of these characters described later). In nearly every case, the Gopher client software will give the users some sort of idea about what type of item this is (by displaying an icon, or a short text message, or the like). The succeeding characters up to the tab form a user display string to be shown to the user for use in selecting this document (or directory) for retrieval.

The characters following the tab, up to the next tab form a selector string that the client software can send to the server to retrieve the document (or obtain the directory listing of the directory). The selector string should mean nothing to the client software; it should never be modified by the client. The last two tab delimited fields denoting the domain-name of the host that has this document (or directory), and the port at which to connect.

In the example, line 1 describes a document the user will see as "About Gopher". To retrieve this document, the client software must send the retrieval string: "Stuff:About internet Gopher" to rawBits.micro.umn.edu at port 70. If the client does this, the server will respond with the contents of the document, terminated by a period on a line by itself.

Similarly to retrieve the directory listing for what the user sees as "Courses, Schedules, Calendars", the client software must send the string "/Courses" to the server. If the item of interest were a search, the client sends the selector string followed by a tab and the search string.

9.3.1 Locating and adding services

While Gopher servers may contain directories and documents, they may also contain references or "links" to directories, documents or other services on other Gopher servers. Some well-known servers also maintain lists of all known Gopher servers (categorized by location or specialty). Most "home" gopher servers are set up with links to such lists of other servers.

The first character of each line in a server-supplied directory listing indicates whether the item is a file (character '0'), a directory (character '1'), or an error (character '3'). While there are a handful more of such type characters, this is all that the Gopher protocol handles intrinsically. It is desirable for clients to be able to use different services and speak different protocols (simple ones such as finger; others such as CSO Name service, or telnet, or X500 directory service) as needs dictate. For example if a server-supplied directory listing marks a certain item with type character '2', then it means that to use this item, the client must speak the CSO protocol. This removes the need to be able to anticipate all future needs include them in the basic internet Gopher protocol; it keeps the basic protocol extremely simple. In spite of this simplicity, the scheme has the capability to expand and change with the times by adding an agreed upon type-character for a new service.

Rather than a proliferation of new Gopher "types" which would entail updates of all clients in the field, Gopher has evolved much faster by deployment of gateway servers. Such servers make services based on other protocols, like ftp, look like Gopher items by translating from the foreign protocol to Gopher.

9.3.2 Using Gopher to publish your own information

Anyone with a desktop computer (or a larger one) and a connection to the Internet can run a Gopher server. With typical Gopher servers, all items below some directory tree become available to Gopher clients on the Internet. Step one is getting the appropriate server software (Table 9.1), a domain-name for your computer, configuring and running your server, and getting your personal directory structure the way you would like. Next, if you would like other folks on the Internet to be able to find your server, you need a link from some other linked Gopher. If your campus already has a main Gopher server, then you should probably request the administrators of that server to make a link to your server. You may also (or alternatively) request that your server be placed on one of the master lists of world-wide Gopher servers. If your server is in Europe, send email to `gopher@ebone.net`; elsewhere send your mail to `gopher@boombox.micro.umn.edu`.

Chapter 10

WorldWideWeb (WWW)

Bebo White

Stanford Linear Accelerator Center
P.O. Box 4349, MailStop 97
Stanford, CA 94309 (USA)
Email: bebo@slac.stanford.edu

10.1 Introduction

The term *hypertext* was coined by T.H. Nelson circa 1965 to describe text (later described as being words, pictures and sound) which is not constrained to be linear (i.e. sequential; Kahn et al, 1988; Nelson, 1990). In particular, hypertext can be viewed as text which contains links to other texts. A *hypertext document* can therefore be thought of as a collection of texts and the links between them which can logically be viewed as a single entity. There is no requirement that the constituent parts of a hypertext document must be stored as a single entity. An *anchor* within a hypertext document is defined as some collection of text which is the source or destination of a link. Figure 10.1 provides a simplistic view of the construction of a hypertext document.

In 1965 hypertext was an interesting philosophical concept. Meaningful applications of these concepts were difficult to conceive. However, in the 1990s high-speed networks and information servers have made hypertext an important concept in the design of information retrieval systems. The definition of network protocols specifically designed to emulate hypertext links have made hypertext documents a reality.

10.2 What is WorldWideWeb?

WorldWideWeb (WWW or W3) has introduced to information retrieval on the Internet, the concept of *hypertext*. Rather than rely upon an existent hierarchy of information or a keyword-based search, this method attempts to link information in a manner which

A. Heck and F. Murtagh (eds.), Intelligent Information Retrieval: The Case of Astronomy and Related Space Sciences, 127–133.

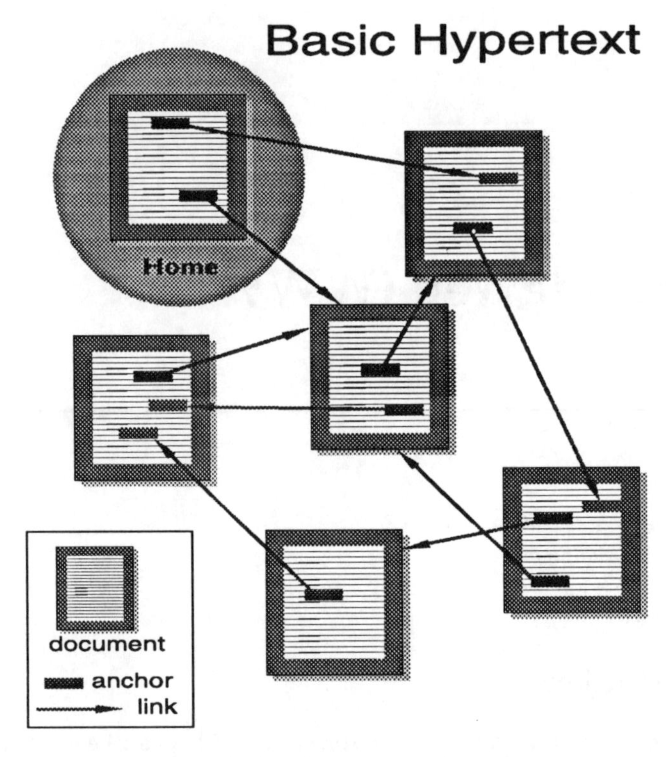

Figure 10.1: An illustration of hypertext.

mimics the human association of ideas. As a result, the associative links, access paths, etc. between information and/or information sources more closely resemble a spider's *web* rather than the conventional tree-like or directory-like structure. The topology of the Internet nodes within this information network can also be thought of as a similar web.

The WWW user is therefore able to follow a multiplicity of paths (even circular) in order to find the information required. In the meantime, such paths may allow the user access to additional similar or dissimilar information, associated information (e.g. definitions, etc.), and access to other information retrieval systems.

10.3 WWW Features

10.3.1 How does WWW look to the network?

Like a number of other information retrieval systems (e.g. Gopher and WAIS), WWW presents network-distributed information. The WWW architecture involves two programs, a client "browser/reader" and a server, communicating across the network. The "information bus" which connects clients consists of a uniquely defined set of standards

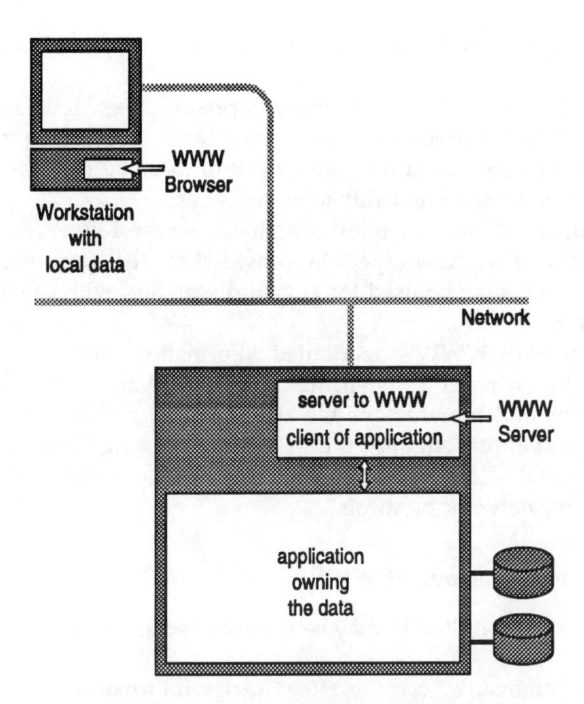

Figure 10.2: The browser and server relationship.

and conventions. The three conventions of this "information bus" are Universal Resource Locators, a set of protocols, and a set of data formats. At this level, the user-perceived "web" is composed of named documents some of which may in actuality be searchable indices. The browser can present the documents and the relationships between them in a manner suited to the needs and/or desires of the user. The server is free to display the documents either by sending real files (either with or without hypertext links) or by generating virtual hypertext dynamically in response to a request to some service available to the server. This is how a *gateway* server can provide a hypertext image of all the data in another system. Therefore, information providers, such as some of the others described in chapters 7, 8 and 9, are not forced to change their interfaces or convert information. This process is handled by the gateway server. Figure 10.2 provides a simplistic view of this browser/network/server relationship.

10.3.2 How does WWW look to the user?

To a user, WWW consists of two fundamental operations – following a hypertext link and performing a keyword indexed database search. The usual WWW client interface (i.e. browser) resembles a menu (commonly called a *page*) which contains active buttons (the hypertext links) or keys. The entry-level page at a particular node is commonly referred to as that node's *front page*. On platforms which allow hypertext-processing (e.g.

mouse-driven systems), the user "points and clicks" on these buttons in order to follow a link.

The default "line-mode" or TTY browser presents these links as numeric menu items. The number of the link to be followed is typed by the user on the command line. Consistent with WWW's hypertext metaphor, none of these buttons/links need be sequential or hierarchical in their relationship to one another.

A link can be a dynamic request to an index server. Keywords for index searches are usually typed in a window especially provided for that purpose. The Midas interface is one such which may be used for keyword searches, with the index server being the Gopher Server.

Consistent with WWW's distributed information model, the links followed from a page may be on many different machines anywhere on the network. As the user pushes buttons and browses one file after another, he/she may be actually querying information servers throughout the network, exploring therefore a large information *web*. The traversal of this web is usually transparent to the user in that WWW handles all of the necessary network protocols.

10.3.3 General features of WWW

The general features of WWW may be summarized as follows (Berners-Lee et al., 1992):

- WWW Can Insure Access to Most Current Information

  Information need only to be represented once, as a reference to the original/master source document may be made instead of making a local copy.

  Links allow the topology of the information to evolve, so modeling the current state of a topic area of interest is possible at any time without constraint.

- WWW Optimizes the Use of an Information Source

  The user-available information web stretches seamlessly from small personal notes on local workstations to large databases anywhere on a network.

  Indices are documents, and so may themselves be found by searches and/or following links. An index is represented to the user by a *cover page* that describes the data indexed and the properties of the search engine.

  The documents in the web do not have to exist as files; they can be *virtual* documents generated by a server in response to a query or document name. They can therefore represent views of databases, or snapshots of changing data (e.g. time-dependent data, jointly-authored works-in-progress).

10.4 WWW Conventions

As described earlier, the success of WWW operation depends upon three conventions – a set of protocols, Universal Resource Locators and a set of data formats.

10.4.1 Protocols

A number of existing protocols, and one new protocol, form the set with which a WWW browser is equipped. Standard protocols used are:

- FTP

  FTP (File Transfer Protocol) allows access to libraries of documents, software, etc. available on ftp servers which allow "anonymous" access.

- NNTP

  NNTP (Network News Transfer Protocol) allows access to Usenet Netnews servers and the display of news groups and articles.

A WWW browser is also equipped to handle two of the protocols of other information retrieval systems discussed in this book:

- WAIS

  Allows gateway access to WAIS (Wide Area Information Servers). WAIS is described in detail in chapter 8.

- Gopher

  Allows gateway access to the Internet Gopher system. Gopher is described in more detail in chapter 9.

A new protocol, HTTP (Hypertext Transport Protocol), was developed to allow document retrieval and index search (Berners-Lee, 1991). This protocol coordinates communication between WWW browsers and servers and is easily implemented in shell script, perl, C or other programming languages. HTTP is designed to be as efficient and compact as possible so as to minimize round trip delays between the browser and server nodes. A great deal of its flexibility lies in the Universal Resource Locator.

10.4.2 Universal Resource Locator

The Universal Resource Locator (URL) is used to define the essential information in a link. In earlier WWW documentation, the URL was referred to as the UDI (Universal Document Identifier), but has been changed to reflect the widening variety of information resources now available. The URL is designed to be compact and printable (i.e. contain no special characters). The URL contains fields identifying the applicable protocol, a server specification (usually just the fully-qualified Internet name), a port address, a document to be identified on that server, and a search to be performed if necessary. It may optionally contain a field which specifies a particular part of a document to be selected (an *anchor*) when the document is presented.

10.4.3 Data formats

For WWW the lowest common denominator for data is plain text. This format allows the ready display of simple documents and text output of index searches.

WWW servers may also generate simple hypertext documents written in a specially-designed text-processing language, *HTML*. HTML (Hypertext Markup Language) is a subset of the popular SGML (Standardized General Markup Language). It has a few simple formatting options (tags) which allow it to be effectively used for on-line documentation as well as menus and search results. HTML permits browser-dependent formatting of hypertext documents and the definition of links. For example, in an HTML document a markup tag such as #anc would contain a URL (Universal Resource Locator) to create a hypertext link.

As formats for representing data continually evolve, access to files of these data types must become generally available. The WWW architecture proposes a negotiation between client and server to agree on a document format for transmission. *Format negotiation* is available for some formats in specific browsers. Formats of particular interest are graphic files (e.g. GIF, TIFF), text-formatted files (e.g. DVI, RTF) and sound files.

10.5 How to Get More Information on WWW

10.5.1 Public access to WWW

A fully implemented version of WWW is available via *telnet* to *info.cern.ch*. No username or password is necessary in order to use this service. This example, since it is via telnet, uses the linemode browser. The user will be able to explore the web, but will not be able to use a hypertext-based interface.

The *front page* of *info.cern.ch* is highly tailored for use by CERN, the European Particle Physics Laboratory in Geneva, Switzerland (though it changes quite often). However, links do provide access to a wide variety of other information sources, including WAIS and Gopher (discussed in chapters 8 and 9). For example, following links entitled *academic information* lead to resources in the areas of astronomy and astrophysics.

10.5.2 Obtaining WWW software

Access to WWW server and client software for a wide variety of platforms is available from numerous sources via anonymous ftp. Using archie (described in chapter 7), a user can identify a convenient ftp site from which to get the software by issuing the command `prog www` or `prog WWW`. Pointers to sources for all of the popular browsers (e.g. Viola, Midas, tkWWW, XMosaic, etc.) are also available. Archie can also be used to point the user to the wealth of WWW documentation that is also available.

The guaranteed latest WWW server software (for a wide variety of platforms) and some browser software (including the linemode browser) is available via anonymous ftp from `info.cern.ch`

10.5.3 Additional information sources

There are presently two electronic mailing lists available to those interested in WWW –

- www-announce

  This list keeps subscribers informed as to WWW progress, new software releases and new data sources.

- www-talk

  This list provides a mechanism for technical discussion for those users interested in developing WWW software, WWW protocol evolution and WWW-related technologies.

In order to subscribe to either of these lists, send electronic mail to `listserv@info.cern.ch`. The body of the mail should contain only the line:

`add www-announce`

or

`add www-talk`

The name and address occurring on the From: line of this mail will be added to the mailing list(s).

References

1. Berners-Lee, T., "HTTP As Implemented in WWW", CERN, December 1991.
2. Berners-Lee, T., Cailliau, R., Groff, J.-F. and Pollermann, B., "World-Wide Web: the information universe", *Electronic Networking*, Meckler, Spring, 1992.
3. Kahn, P.D., Pau, Meyrowitz, "Guide, HyperCard, and Intermedia: a comparison of hypertext/hypermedia systems", IRIS Technical Report 88-7, Brown University, Providence, RI, 1988.
4. Nelson, T., *Literary Machines*, 90.1, Mindful Press, Sausalito, CA, 1990.

Chapter 11

Information in Astronomy: Tackling the Heterogeneity Factor

Miguel A. Albrecht

European Southern Observatory
Karl-Schwarzschild-Str., 2
DW-8046 Garching/Munich (Germany)
Email: malbrech@eso.org, eso::malbrech

and

Daniel Egret

CDS, Observatoire Astronomique
11, rue de l'Université
F-67000 Strasbourg (France)
Email: egret@simbad.u-strasbg.fr, simbad::egret

11.1 Introduction

Over the past two decades, the increasing efficiency of cameras and photon collecting devices used by astronomers has generated an unprecedented accumulation of data. This phenomenon, a characteristic of the so called *information era*, has not been accompanied by the corresponding development of tools and systems to handle, analyse, store, retrieve and disseminate observational data. The result being that only a small part of the data is immediately used and published, in the investigations for which it was planned. At the same time, the always larger investments involved in modern telescopes or space missions have raised the question of cost-to-benefit ratio in the management of observing facilities – in particular when it is rather usual to find a large pressure factor on the usage

A. Heck and F. Murtagh (eds.), Intelligent Information Retrieval: The Case of Astronomy and Related Space Sciences, 135–152.

of these facilities (e.g. for HST and the ESO NTT there are about four times more observing proposals than time available).

Another aspect which has contributed to a drastic change in the way we perceive data, is a noticeable trend towards multi-wavelength studies: i.e. studies of objects or phenomena in several of the classical energy ranges (optical, infrared, radio, etc.) in parallel. This science approach has raised the need to be able to access data holdings from different instruments and to be able to bring these data, or information related to it, into a coherent presentation.

The answer to this questions is a wealth of activities aiming at archiving, dissemination and cross-correlation of astronomical data that have evolved in the last years. The pace of development in this area is in fact very fast, with new systems being funded and new projects taking off every year. The most important to mention here are (a) the adoption by the ESA Science Programme Committee of an archival policy that foresees the creation and operation of science archives for all upcoming ESA science missions (May 1992); (b) the public release of both the Astrophysics Data System (ADS) and the European Space Information System (ESIS) (spring 1993); (c) a number of conferences and workshops dealing with this topic (Heck and Murtagh, 1992; Hanisch et al., 1993; and Albrecht and Pasian, 1993). An overview of activities in this area has been given by Albrecht and Egret (1991).

In this paper we describe first the current usage of archives and on–line data facilities. We discuss thereafter, aspects involved in retrieving information from data systems and associated practical problems. In section 11.3 we give a model aimed at understanding the various levels of heterogeneity to be overcome when attempting multi-wavelength studies and describe some efforts from the data system praxis. Finally, a perspective is drawn towards the next generation of *information system* in astronomy.

11.2 Archives and Databases as Research Tools

A first large scale proof of the utility and value of astronomical data archives came with the decision of NASA, ESA and SERC to set up an archive of images and of reduced spectra obtained by the International Ultraviolet Explorer satellite (IUE), and to service archive data requests from their respective science community. As shown by Wamsteker (1991), this approach is highly successful and has certainly allowed the user community to grow far beyond the original IUE observers. Usage statistics show that, in the average we are close to the rate of three spectra dearchived for one spectrum observed. To say it differently: the scientific result with the existing archive and 14 years of IUE observations is equivalent to 56 years of IUE observation without an organized archive.

A new dimension is currently being explored as HST has started producing results. Here, the aims are to cope with very large quantities of data (of the order of Terabyte/year) as well as to maximize scientific throughput. Current statistics from the HST-archive operated by the European Coordinating Facility (ST-ECF) show that while HST is acquiring an average of 1,200 scientific observations per month (summer 1992) the ST-ECF archive is delivering an average of 400 observations per month to the community (Pirenne et al.,

1992); and while the first figure is expected to remain constant as time passes, the number of observations that are released into the public domain increases steadily.

In the domain of Solar observations, the example of the SMM satellite (Solar Maximum Mission) shows the role of an archiving organized early, at the Operations Center itself. The satellite was launched in 1979; the instruments failed one year later and were repaired in 1984; the mission ended in 1989. An analysis of the roughly 700 papers published in the period 1979-89 illustrates that the scientific work could go on actively, based on the archives, during the period when the satellite stopped producing data.

The La Palma archive stores regularly observations performed with the Isaac Newton Group (ING) of telescopes at La Palma, Canary Islands, and with the Westerbork Synthesis Radio Telescope (WSRT) in the Netherlands, for already a number of years. Data request statistics (Raimond, 1992) show that on the average, around 1,000 observations are retrieved from the archive per year (optical telescopes only) compared to about 11,000 new observations made. The lower utilisation figure might be due to the fact that the ING telescopes archive raw (unreduced) data plus calibration frames, and so, archive researchers must reduce the data by themselves before actually being able to decide whether or not a particular frame is suitable for a certain purpose.

Systematic archiving is also well underway at the European Southern Observatory (ESO) with the New Technology Telescope (NTT). The archive holds, in January 1993, about 13,000 observations of which, to this time, only about 6,000 have been released into the public domain. As of September 1992, also data acquired at the Canada-France-Hawaii Telescope (CFHT) is being archived on a routine basis (Crabtree, 1992). The archive istelf is located at the Canadian Data Center (CADC, Dominion Astronomical Observatory).

The amount of data coming out of the ESO/NTT is estimated to be between 0.2 and 0.4 Gbytes/day, i.e. of the order of 100 Gbytes/year. This amount will grow rapidly: the usage of 2048×2048 CCD frames (instead of 1024×1024 presently) is scheduled for the near future (Albrecht and Grosbøl, 1992). This trend towards larger and larger volumes of data makes it extremely difficult for individual researchers to keep up with their specialities, in terms of being aware of new data as well as tracking new ideas. In response to this problem, attempts have been made to build databases of *results* of scientific research. A pioneer in this area has been the Centre de Données astronomiques de Strasbourg with its Set of Identifications, Measurements and Bibliography for Astronomical Data (SIMBAD; see Egret et al., 1991). Here, one would find all "published knowledge" on a particular astronomical object, for instance its spectral class (for stars), magnitude, distance, etc., as well as references to articles dealing with the object. Other activities have developed, partly independently, in the same direction: the EXOSAT and HEASARC databases for high energy data, and the NASA/IPAC Extragalactic Database are two primary examples (see White and Giommi, 1991, and Helou et al., 1991, respectively).

The usage statistics of all these facilities show how much they are appreciated by the community: 900 *user accounts*, from 300 institutes in 34 countries for SIMBAD, generating around 2,000 remote logins per month. The number of logins for NED is also about 2,000 per month, while EXOSAT (which, of course, concerns a smaller specific community), announced a total of 3,261 sessions for its first year in 1990. STARCAT reports about

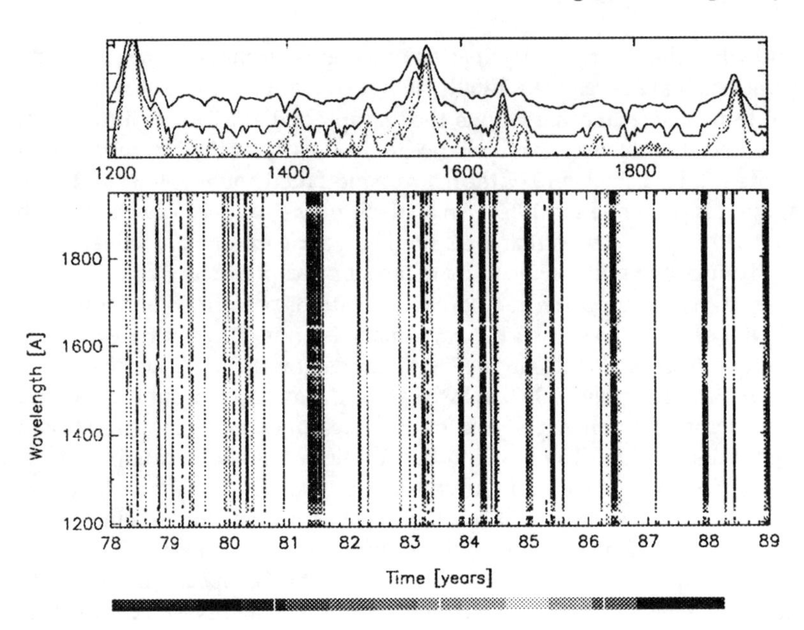

Figure 11.1: Spectra observed by IUE of NGC 4151 over the period 1978–1989. The curves on top of the graph show averages of the spectra over periods of time where the spectra didn't change noticeably (within dotted lines. Courtesy of P. Giommi, ESIS correlation environment; see Giommi et al., 1992).

1,900 query sessions per month (Pirenne, 1993).

Altogether these figures show that archive retrieval is steadily becoming an integral part of scientific research. As an example, Figure 11.1 shows the unparalleled quality of a long term archive with homogeneous data contents: spectra taken by IUE of NGC 4151 over the years 1978-1989.

In the area of auxiliary information, such as e.g. opacity coefficients for stellar atmospheres, independent groups have adopted different approaches in regard to data management: the Livermore group tends to prioritize the re-computation of data on request, without systematic archiving of the results; while the "Opacity" project has chosen to store computed tables of atomic data in a database which can be used directly by interested astronomers or physicists (the TOPBASE database now available on-line through the CDS in Strasbourg). The increasing rate of information production also affects scientific journals: it becomes more and more difficult for editors to cope with the publication of extensive tables. *Electronic publishing* is now being seriously considered (see the special supplement of AAS Newsletter dedicated in November 1992 to this question; see also

Heck, 1992). Starting in 1993, the AAS will publish twice a year CD-ROMs containing tabular data accompanying articles published either in the *Astronomical Journal* or in the *Astrophysical Journal*. At the same time, *Astronomy and Astrophysics* announces that tabular data published in *Astronomy and Astrophysics Supplement Series* will be made available on-line through the anonymous ftp facility at the CDS. In order to be able to index and catalogue the information in a useful way, the AAS plans the use of markup languages (SGML strictly speaking; and LaTeX when used following specific rules). This will allow the application of Intelligent Information Retrieval Systems for navigating within the databases of electronically stored publications.

11.3 Retrieving Information

Accessing data archives becomes a crucial issue as soon as it is realized that retrieving information from archives will not remain the privilege of a few specialized *archivists*, but is opened to a wide scientific community, and in practice made available to any astronomer from his/her desktop terminal.

In order to better describe the processes involved in information retrieval, we make use of the model illustrated in Figure 11.2.

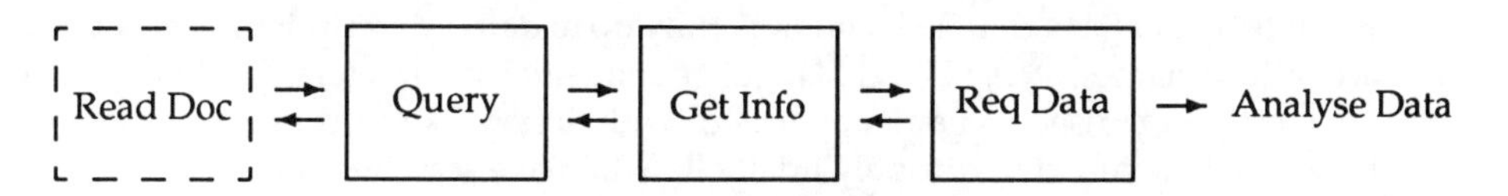

Figure 11.2: The information retrieval paradigm

In the following we use the term *system* to denote a general data collection that has an *access method*, an *index* of some kind and consistent *nomenclature*; thus it applies to both computer- and hardcopy-based data sets. We include the query language of an on-line system as part of its access method.

In a "natural" sequence of events, one would start by reading the documentation of the system in question, find out what kind of data it holds, how the index is structured and so on. The next step would then be to search through the system to obtain information about observations (spectra, images, etc.). This information is eventually used to carry out the actual retrieving of those data sets to be analysed. This model, even if rather simple, illustrates the distinction between the data holdings, often called archives, and the system used to describe and index the archive contents, also called catalogue, log or index. In an on-line facility the latter constitutes the *information retrieval* (IR) system. The difference is that while archives deal with data (observations or derived products), the IR system deals with *descriptions* of data that logically make reference to actual data sets.

In the following we will summarise standard issues and mention some typical problems associated with each of the steps described in our model. This article does not attempt to be comprehensive, and examples are given only for the purpose of illustrating a particular problem or feature. Detailed descriptions of a number of data archives and databases in astronomy can be found in Albrecht and Egret (1991).

11.3.1 Documentation

Most people skip the documentation and go directly to query the system, i.e. to find out what data can be used straight away. This has some unfortunate drawbacks, because managing to use the system successfully without ever consulting a user manual (e.g. to formulate a query in the local query language or to find the correct card in the box), does not mean that the results will be interpreted correctly. This is due, as we will see later, to the lack of standardisation on naming conventions, semantic meanings and physical assumptions across different systems. The documentation is in general the only source where these conventions and formats are described.

Most systems will provide two kinds of documentation:

what i.e. explanations of the contents and structure of the archive, including physical units and coordinate system used, assumptions made during its creation, etc. In the case of observational data, the documentation will usually make reference to, or include descriptions of the instrumentation used.

how i.e. instructions regarding the usage of the system in question. On-line systems will mostly deliver a detailed (and hopefully up to date!) description of its network access (usernames, etc.), its query language and search parameters. Off-line systems are mostly organised as hardcopy lists or with library-like card indexes; therefore they will generally describe only briefly how to make searches, if at all.

There is hardly a way to standardise the documentation of IR systems considering the wide variety of data kinds and system functionality that these may include. However, it would be highly desirable to promote the establishment of a *style guide* of archive/catalogue documentation, which would address typical problems arising from making data available to the community at large.

One of the often underestimated problems is the question of *unspelled assumptions*: most systems have been developed with a well defined and often closed user community in mind. The result is a number of undocumented assumptions that are supposed to be "well known" and thereby incorporate the danger that information might be misunderstood by users outside the original user group. The descriptions are also frequently biased by the environment in which the system was developed, e.g. VAX-VMS-like command syntax or Unix-like case sensitivity that are assumed not to require further explanation. It is not unfrequent for a user to be stuck somewhere without knowing how to leave the system – unless unplugging the computer.

More generally, the quality of a user manual or data description is correlated to the resources that have been dedicated to its writing, i.e. time spent, expertise and linguistic capabilities of the writer, etc. Only a few projects have in the past made a serious effort to document the systems they created. In fact, the rule is that archive projects concentrate effort on the actual creation and setup of the system and consider documentation only as a second-order task. Only a partial answer to this concern will come with the efforts put into the development of more and more sophisticated graphical user interfaces (hopefully providing easy access to on-line helps).

11.3.2 Query formulation

A considerable part of the effort spent in querying a system is used up in learning how to formulate the query according to the syntax of the specific query language: this may imply local specialities like coding schemes for object classes or spelling rules for object names. There is a variety of on-line systems available world-wide to the astronomical community and, owing to historical reasons, most projects have taken an ad-hoc approach in the set-up and development of data systems, with the result that astronomers are confronted with a wealth of different user interfaces (Pasian and Richmond, 1991).

The query formulation is generally supported by a (more or less) sophisticated user interface which carries part of the knowledge about the database or archive architecture. Typical features are:

- Most systems are *command driven*, i.e. the interaction paradigm is that of "prompt and answer", having the possibility of invoking a help command if needed. Examples: IUE, ARCQUERY (La Palma), SIMBAD, etc.
- Some systems also support "screen" oriented interaction such as menus, forms (masks) or lists. Examples: STARCAT (Pirenne et al., 1992) and NED (Helou et al., 1991).
- A few systems support "batch" processing in addition to interactive sessions. The La Palma/ING/WRST Archive also supports remote interaction via email (Raimond, 1991); this feature is also recently available for SIMBAD and STARCAT (STARMAIL, see Pirenne and Albrecht, 1993).

With reference to search capabilities, the user will find that most on-line systems have indexes on RA and DEC. Archives of observations will, in general, also provide indices on some other relevant parameters like time and date and name of observer. This is an important (although often not explicitly mentioned) feature of the system, with obvious consequences on the response time of any generic query.

The nature of data acquisition in astronomy implies a specific consideration of the organization according to the targets (i.e. the astronomical objects). Unfortunately, the nomenclature of objects is a tough problem, and only a few systems deal properly with object names. This includes:

1. homogenisation of object names throughout the database (see for instance Egret et al., 1992);
2. handling of synonyms and alternate names;
3. handling of alternate spellings (e.g. case sensitivity).

In fact, one is strongly discouraged to search *any* archive for object names. Coordinate searches should be preferred whenever possible. Noticeable exceptions are SIMBAD, and NED for extragalactic objects, which are the only databases known to the authors capable of resolving name conflicts in an appropriate way.

11.3.3 Obtaining information about the data

At the end of a successful query session, the user will have a number of descriptions, each corresponding to a particular data set. It is important to emphasize at this point that a query will retrieve *information about the data*, also called *metadata*, and not the data sets themselves.

In most cases the first search will deliver much more information than is needed, e.g. when looking for all objects of a particular type in SIMBAD or specifying a large sky segment in the Guide Star Catalogue (Jenkner, 1991).

This is a well known problem in Information Sciences and the process that follows a *broad query* is often referenced as "query refinement": incrementally added constraints to the query, also called *qualifications*, will reduce the query set until the number of items left out represents a reasonable amount of information to be analysed. The amount will vary strongly with the kind of data with which one is dealing. For instance, when querying a bibliographical data bank (Rey-Watson, 1991), a quick look at the abstracts of the selected papers will already reveal their usefulness, and so a query sample of the order of some tens will still be manageable.

The ability to discriminate between useful and not useful items in a particular context is directly related to how much information the user must browse through before he/she can decide to keep or discard it. Ideally, a system should allow users to design dynamically the display of results, even allowing the generation of "summary" fields that are built out of the combination of database fields (e.g. flux and exposure time). Unfortunately, only few systems have been designed to cope with this need, compelling users to filter out the relevant bits from screenfuls of information.

It is important to be aware that some systems may not check at all the information they incorporate into the database, mainly because of lack of resources. In fact, quality control is the one task most resource-intensive in the operation of information systems.

11.3.4 Requesting or retrieving data

Once some data sets have been identified as of interest in a particular project, the next step will aim at getting hold of the actual data. Most systems will support the off-line retrieval of data, i.e. the mailing of tapes containing the specified files. A few systems will also support on-line retrieval possibly restricted to the on-site local area network. The data format in most widespread use by existing archives is FITS (see Grosbøl, 1991, for a recent review).

The next question is whether or not the data obtained from an archive will be already calibrated. Some systems – mainly space projects – will deliver data in physical units, reduced via a standard calibration pipeline. The IUE Uniform Low Dispersion Archive

(ULDA) (Wamsteker, 1991), represents a notable example where a large effort was put into building an homogeneous and easily accessible data archive. The EXOSAT database (White and Giommi, 1991) offers on the other hand access to "results" rather than only reduced data; i.e. data sets derived from the scientific analysis of the original observations.

Finally, in order to make a scientific analysis of an observation properly, one would ideally need to find in the description (or *header*) all observational parameters, even if they do not directly influence the physical format of the data set. Experience shows that more important than the way chosen to describe observations, is whether or not the appropriate data analysis package will be able to interpret the various keywords correctly, in particular, non-standard FITS keywords. In fact, while a format like FITS does provide a standard mechanism to describe *data* in a basic form, it does not provide a standard way to describe *observations* as a whole. This is a key issue to consider when planning to analyse data from yet unknown observing facilities. Ideally, one would like the data from equivalent observations (e.g. from two optical telescopes of the same class) to be comparable. This requires that the description of the observations is understood by the data analysis packages that are used to do the comparison. Currently, there is hardly any compatibility between systems accross observatories and analysis software, however, a number of tools are being developed to tackle this problem (see also section 11.5).

When retrieving bibliographical references, there is no standard format equivalent to FITS for data. A markup language like SGML or LaTeX would provide an answer to this problem. However, this would require the development of appropriate tools and facilities that do not exist yet in the astronomical domain.

A fundamental issue yet to be clarified is the feasibility of comparing data obtained with different instrumentation. Most probably there is no general answer. A different case is given when cross correlating results rather than observational data directly. An example of the latter is shown in Figure 11.3 where an EXOSAT image has been superimposed with known IRAS point sources in the same region of the sky (the Pleiades).

11.4 The Heterogeneous Factor

In general, we observe that systems can show differences with regard to other systems at various levels. We have summarised these levels in Table 11.4.

physical	data formats, protocols, access methods, etc.
syntactical	coordinate systems, naming conventions, spelling, etc.
semantical	keyword meaning, classifications, etc.

Table 11.1: The heterogeneity factor. Levels at which IR systems show differences with regard to others.

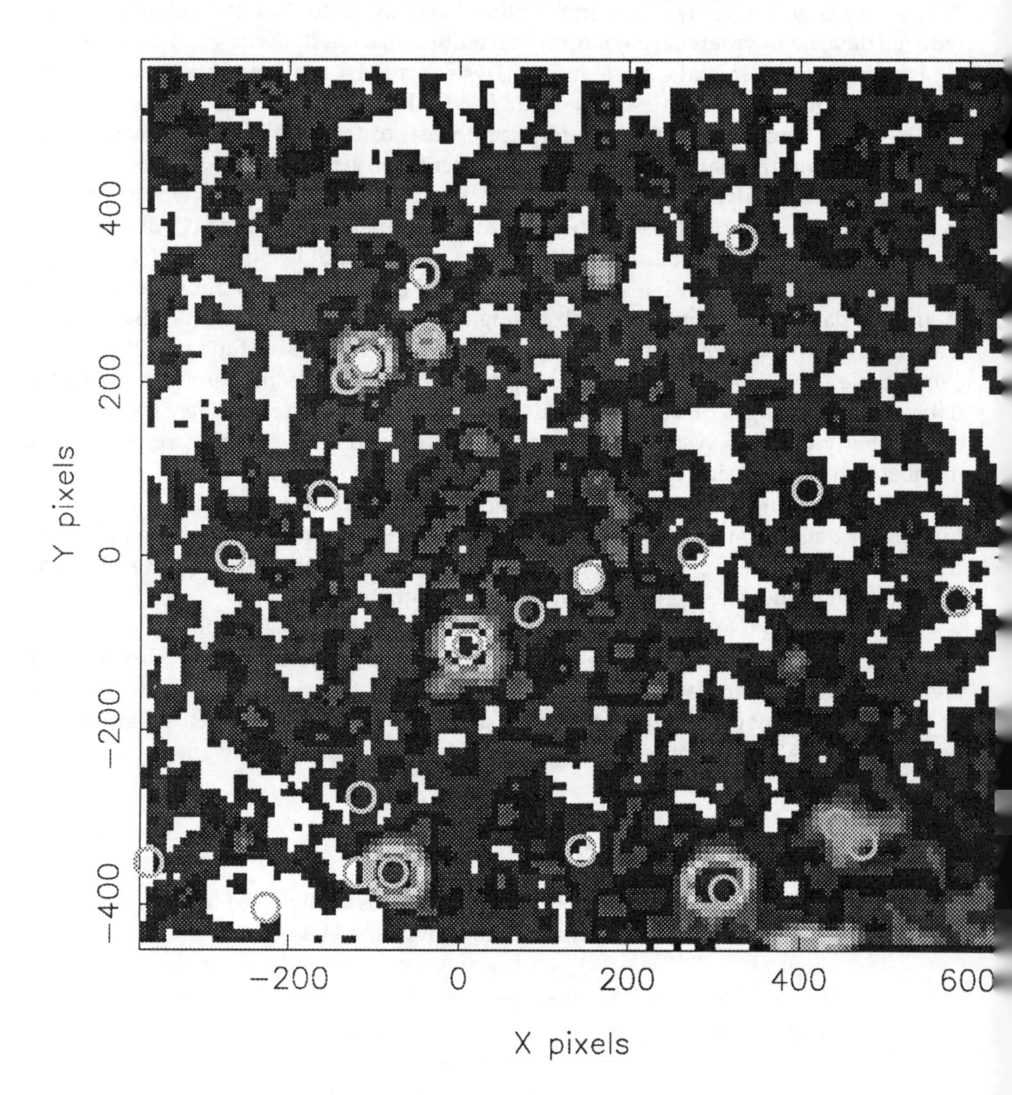

Figure 11.3: The Pleiades as seen by the CMA instrument of the EXOSAT satellite. Known IRAS point sources are superimposed as circles (courtesy of P. Giommi, see Giommi et al., 1992)

In the following we will describe some successful strategies implemented at various data systems that aim at solving heterogeneity.

11.4.1 Identifying astronomical objects

Varying nomenclature and ambiguities on object names are an obstacle when attempting to retrieve observation records. For instance, there may be up to several dozen names for bright stars: constellation name; names in reference catalogues (HD, HR, BD); variable or double star name; names in observation catalogues such as SAO, or IRAS; etc.

Let us take the example of the log of observations of the IUE satellite: this catalogue is the basic index for retrieving (de-archiving) the actual spectra. It contains a string field for object name, freely (and sometimes in a fanciful way) filled by the observer; there is no specific rule for uniquely naming an object. The only way then, of ensuring that all observations of a particular object are selected is to access the log via the coordinates field. This field, however, is also filled by the observer during the observations: any error in the name and/or coordinates of the object has the consequence of putting the corresponding archived image into a "black hole" where the standard procedures of de-archiving are unable to find it! Several hundreds of images (i.e. one or two months of IUE observations) would simply have been lost, without a specific effort for homogenizing the nomenclature within the IUE archive. For this reason, the IUE project started, in collaboration with the CDS, the homogenization of the IUE observations log as a step in the preparation of first the ULDA Archive, and now the IUE Final Archive. For that purpose, entries in the IUE log were cross-matched against the SIMBAD database. A "unique" identifier, following a pre-defined list of priorities, was determined in this way for each entry in the log (Egret et al., 1992). This identifier is then efficiently used as a key index in the homogenized IUE archives.

The experience gained at the CDS during this task has been used to populate a small knowledge base containing the rules needed for translating a name written in a free format into the standard syntax used up to now within SIMBAD. These sets of rules have now been reused by the *Sesame* module which has been included in the Simbad DBMS in December 1992 (see more about Sesame below). An example: "RR Lyrae" is translated into "V* RR LYR" where V* is the standard abbreviation for the catalogue of variable stars in SIMBAD, and LYR the standard abbreviation for the constellation name. The NASA/IPAC Extragalactic Database (NED) maintains, since its first release in 1990, a similar (but larger) set of rules managed by the so-called "egret" module, based on the "lex" lexical processor generator (see Helou et al., 1991). Here also the idea is to interpret the object name given in "natural language" and to translate it into a standard internal format.

A different approach is currently being implemented at the ESO archiving facility. Instead of translating user-given object identifiers, the selection of objects is based on their coordinates. This is possible because the telescope pointings (coordinates) are written for every observation by the telescope control system without user intervention. The user interface (STARCAT), is provided with a "name resolver" function that makes use of a transparent connection to *Sesame* (SIMBAD) in order to retrieve the coordinates of a particular object. The actual query is then formulated in terms of a cone projected onto the sky centered on the coordinates obtained via SIMBAD with a radius which corresponds to an error estimation on the coordinates field. The same function has been successfully implemented in the first version of ESIS. A function called "query by name" allows users to enter an object name which is resolved into coordinates via SIMBAD before the query is executed in the ESIS Cats & Logs database.

11.4.2 Describing observations

One major problem in archiving observational data arises from the need to describe all instrumental parameters that make both the static and dynamical configuration. For large instruments, the number of parameters to be coded can easily exceed 300. For orbiting telescopes, these parameters may be augmented by telemetry and engineering information concerning the telescope position and status. Ideally, one would like to have all instrumental parameters written with the data in the FITS header. The basic FITS format, however, does not include a convention on how these parameters are to be named, nor what units are to be used. Moreover, FITS keywords can only be 8-characters long, so that the coding of a few hundred names makes keywords become extremely cryptic. Several solutions to this problem have been adopted – mostly ad-hoc – by different projects, resulting in total incompatibility.

The ESO Archive facility has taken initiative to set a standard to describe observations making use of the `HIERARCH` FITS keywords. In this notation, the e.g. filter wheel position on the red arm of an instrument is described

`HIERARCH INS REDARM FILTER =' #601'`

The `HIERARCH` keyword itself, not being an adopted FITS standard, does not harm the data analysis software which by convention simply ignores any header record that does not have a "=" character in the 9th column. By introducing such notation, the logical transport of information accross instrumental configurations becomes feasible – provided the data analysis systems support it.

11.4.3 Structured information

Whenever information is to be ingested into, or extracted from, a database, the problem arises of how to describe the different elements that make part of an information item (e.g. columns of a table). In textual documents these might be sections, graphs, title, authors, etc. In the case of astronomical catalogues, or more generally, tables of numerical data, one needs to refer to bibliographical references, notes, explanations on physical units and numerical procedures, etc. In the case of observational data one needs to describe the instrumental configuration used. One possible solution to this problem is the application of a language that expresses information structure: a *markup* language. The basic idea behind this approach is that each single element in the information stream is "tagged" with its field name, and that processing software take special actions when fields are encountered while processing the data. The action taken on a particular case depends, of course, on the purpose of the processing task. The simplest example is that of an astronomical catalog which includes its title tagged as `title`. Then, a typesetting task can print the title in a larger font size, possibly in bold face, whereas the task ingesting the catalogue into a database will put the contents of the title in a particular database field. Again, a catalogue cross-correlating task might use the title contents in a different way or simply ignore it. The input source in all cases is the same: a markup document.

Although there exists a standard for markup (Structured Generalised Markup Language, SGML: van Herwijnen, 1990), the lack of tools and processing software for this language have largely impeded its widespread establishment in the information market. In the following, we will mention an example of the use of markup in connection with database interface that is implemented using the LaTeX language syntax (Lamport, 1986). Since the Spring of 1992, ESO accepts the submission of proposals for observing time at the La Silla Observatory via electronic mail. The proposal format consists of a number of fields to be given by the author in a document with the standard LaTeX syntax. Each of the fields has a name and a particular meaning, e.g. `\Telescope{...}`, `\Instrument{...}`, etc. Upon reception of the proposal at ESO, an automatic task first checks the consistency of the proposal (some fields may require the specification of subfields, etc.) and starts, for passed entries, both one typesetting task and a database ingest task that includes the proposal in the observing programme database. The use of the LaTeX syntax was preferred due to its wide distribution among the science community. The typesetting task is nothing other than running LaTeX on the text with the inclusion of a number of macro definitions that reproduce the appearance of an observing time proposal form on the printed paper (which is then passed on to the peer review committee). The database ingest task consists of a specially designed parser that interacts with the DBMS engine to ingest the relevant information. Once the proposal is in the database, it can be extracted, again as a LaTeX markup, if further processing is required.

11.4.4 Correlation environments

The challenge of information systems lies in providing both transparent access to the information contained in distributed archives and, in particular, in returning results in a useful form. The latter is the realm of "correlating environments". It is important to note that while the comparison of observational data from different instruments will remain a very difficult task (more due to different instrumental characteristics than due to different file formats), the comparison of results, i.e. knowledge extracted from the data, can be easily accomplished across spectral domains and data sets. A good example is given by the cross-correlation of point sources identified in the IRAS data with EXOSAT maps (see Figure 11.3), where the comparison of the original data sets would have been simply impossible. In this respect, the total lack of standards for the format and description of observations as a whole that include instrumental details is one the major problems to be overcome.

Figure 11.3 was obtained with the ESIS correlation environment by cross-matching a catalog of sources against an image of a sky region. Another tool worth mentioning here is the ESO-MIDAS table system which offers the possibility to cross-match two catalogues, or any two numerical tables, specifying a "fuzzy" match condition, i.e. a match with an accuracy estimate that selects items that are *close* to the items in the other table (Péron et al., 1992). This feature makes this tool ideal for cross-identifying objects in several catalogues.

11.5 Perspective: Distributed Information Servers?

In previous sections we have shown that the analysis of data spanning several wavelength domains or longer periods of time, that originate from different sources, can generate substantial work, both when finding out relevant information as well as when bringing eventually retrieved data to a shape that makes them comparable. In order to aid the scientist needing to solve an astrophysical question with data from multiple archives, systems were conceived that would overcome heterogeneity and integrate into one single environment: (a) databases as a whole, (b) information services (e.g. directories of astronomical institutions, databases of instrumental characteristics, etc) and (c) end-user owned data collections – so called *personal databases*. The integrating layer would become highly sophisticated, but the return value would promise both capabilities to conduct multi-spectral research as well as a to handle an always larger number of different data sets. These are the ambitious goals of such systems as the Astrophysics Data System (ADS from NASA) and the European Space Information System (ESIS from ESA; see Weiss and Good 1991; Giommi et al., 1992; Heck et al., 1992).

Whether or not the approach taken in large projects like ESIS or ADS will succeed remains to be seen. At the time of writing (January 1993) both systems start to be available to the community at large. There are indeed plenty of reasons for optimism. Even if the current versions of the programs do not include full functionality, first tests tend to indicate their usefulness. However, the experience of building such systems shows that a number of problems may not be as easy to solve as originally planned: (a) the coordination and interface management of a large number of information sources (databases, archives) introduce a large management load that grows with the number of systems attached to the information system; this load has technical, organisational and human components, underestimation of any of these aspects can lead to failure; (b) flexibility towards new technologies decreases with system size; large systems are often surpassed in functionality by smaller prototypes that focus on specific aspects; (c) the evolution of science requirements on information systems is faster than the system's capability to support them; the requirements on system flexibility are orthogonally opposed to the operation of a stable system: this makes any large system operation a permanent source of conflict. It is the same experience that also makes clear that there is a *necessary* rôle for an entity which is independent of project oriented archives and database facilities: an entity capable of fostering research based on multi-mission/project data.

A different approach starts to show up more recently: the appearance of information *servers*. In fact, a conglomerate of services are now available on the network. The general idea is based on a nowadays rather widespread technique known as the client/server approach. The underlying principle states that it is possible to separate a particular processing task into a part close to the user or the requester of a service (the client) and a part close to resources or service provider (the server). In general, the two parts need not reside on the same computer; in fact the approach was developed to support distributed applications. The client requests a service from the server in a well defined protocol over the network. The server may, itself, require services from other servers to fulfill its task.

The client/server approach seems very promising for fostering a more efficient access of databases through the networks. The client resides on the user side (on the user's workstation or mainframe computer) and generates queries onto the network, while the server resides on the database server and is able to receive the queries and send the answer back.The exchange of information through the network can be limited to coded queries, in one direction, and data (possibly compressed) in the other. It is thus possible to keep on the user side all the functions regarding the Graphical User Interface or the on-line help, and take full advantage of modern window technology (GUI). Both ADS and ESIS intend to deliver users with a client software to be installed at the user side.

We mention here some examples:

Sesame: A prototype developed by CDS for accessing the Simbad database: the client, installed on a remote machine (say, a workstation in Garching), generates requests belonging to a pre-defined set of queries (`aliases-of`, `coordinates-of`, `bibliographical-references-of`). These requests are sent through the network, processed on the Simbad server by a Unix script which opens a Simbad session and retrieves the corresponding information; the answer is then sent back to the user, which receives it on his/her screen. This is especially useful when the user already knows what he/she wants to retrieve (i.e. there is no interactivity needed). When used as a function within a program this procedure allows to directly input the answer into the next step of the program.

WAIS: The Wide Area Information Server (WAIS) project started by Thinking Machines Corp. (see chapter 8 for technical details). In summary, the WAIS project defined a query/answer protocol between WAIS clients and servers and has developed public domain software for both entities. A WAIS server provides, in general, access to large document (textual) collections: the most important for astronomy being the one operated at NASA/GSFC which includes the NASA/RECON abstract service.

StarServer: A prototype network access to STARCAT, mainly intended to serve as a guide star database for ESO telescopes. This service provides access to all catalogues included in STARCAT (GSC, PPM, etc.).

Of course, each server has its particular *protocol* which is specific to the task the server executes. Therefore, rather than solving heterogeneity over all layers, they provide one solution for one particular purpose, leaving the integration task to the general working environment at the user side (window manager, local operating system and analysis tools).

11.6 Conclusions

From usage statistics we observe that the more "knowledge" is put into the database, the more appealing it appears to its users. Knowledge can be added to a data system by:

- augmenting its scientific contents, e.g. by implementing a standard pipeline reduction procedure if it contains only raw data, and by producing derived data products, like "quick-look" preview data (Pirenne et al., 1992), phenomenological classification of observations, etc.;
- establishing links between the data sets included in the archive with other informations sources, like e.g. bibliographical references or observations in other spectral domains;
- adding intelligence to the user interface, i.e. providing the interface with astronomically oriented support functions like the capability to search the observations or objects within a projected cone in the sky.

Only a proper balance of these three aspects can achieve best results. Users will not profit from a sophisticated user interface if the data contents are not equally developed, and at the same time, the best database will remain unused if a clumsy user interface hampers its access.

System developers should bear in mind that the mandate of any archive is to keep its validity in the long term; therefore, no efforts should be saved in documenting the system in accurate detail. This is the only way to ensure that data will remain useful even after projects terminate funding level and the associated human expertise evaporates.

With computer technology evolving rapidly, the cost associated with the set up of data archives decreases visibly. This factor, together with a growing awareness among astronomers and those who build science policies of the potential value of data archives will encourage a wealth of new systems to be developed.

The general trend towards distributed computing (client/server approach) is now well established and will strongly condition any new developments. A wealth of "clients" will populate our workstations. Whether or not this will generate a jungle of incompatible special purpose applications will be decided upon on the basis of whether or not standards establish themselves at all levels: from windowing system to data description.

References

1. Albrecht, M.A. and Egret, D., eds., *Databases and On-line Data in Astronomy*, Kluwer Academic Publishers, Dordrecht, 1991.

2. Albrecht, M.A. and Grosbøl, P., "The ESO archiving facility: the 1st year of archive operations", in *Astronomy from Large Databases II*, A. Heck and F. Murtagh, eds., European Southern Observatory, Munich, 169–172, 1992.

3. Albrecht, M.A. and Pasian, F., eds., *Handling and Archiving Data from Ground-based Telescopes*, ESO Conf. and Workshop Proceedings, European Southern Observatory, Munich, 1993, in press.

4. Crabtree, D., presentation to CDS Council, Strasbourg, 1992.

5. Durand, D., Crabtree, D.R., Christian, C. and Glaspey, J., "The CFHT archive system", *Astronomical Data Analysis Software and Systems I*, eds. D.M. Worrall, C. Biemesderfer, and J. Barnes, Astronomical Society of the Pacific, San Francisco, CA, 72–76, 1992.

6. Egret, D, Wenger, M, Dubois, P., "The SIMBAD astronomical database", in *Databases and On-line Data in Astronomy*, M.A. Albrecht and D. Egret, eds., Kluwer Academic Publishers, Dordrecht, 79–88, 1991.

7. Egret, D., Jasniewicz, G., Barylak, M. and Wamsteker, W., "Homogenization of the nomenclature in the IUE log of observations", in *Astronomy from Large Databases II*, A. Heck and F. Murtagh, eds., European Southern Observatory, Munich, 265–270, 1992.

8. Grosbøl, P., "The FITS data format", in *Databases and On-line Data in Astronomy*, M.A. Albrecht and D. Egret, eds., Kluwer Academic Publishers, Dordrecht, 253–257, 1991.

9. Hanisch, R.J., Brissenden, R.J.V. and Barnes, J., eds., *Astronomical Data Analysis Software and Systems II*, Astronomical Society of the Pacific Conference Series, 1993, in press.

10. Heck, A. and Murtagh, F., eds., *Astronomy from Large Databases II*, European Southern Observatory, Munich, 1992.

11. Helou, G., Madore, B.F., Schmitz, M., Bicay, M.D., Wu, X., Bennett, J., "The NASA/IPAC Extragalactic Database", in *Databases and On-line Data in Astronomy*, M.A. Albrecht and D. Egret, eds., Kluwer Academic Publishers, Dordrecht, 89–106, 1991.

12. Giommi, P. et al., "The European Space Information System", in *Astronomy from Large Databases II*, A. Heck and F. Murtagh, eds., European Southern Observatory, Munich, 289, 1992.

13. Heck, A., Ciarlo, A. and Stokke, H., "StarWays: a database of astronomy, space sciences and related organizations of the world", in *Astronomy from Large Databases II*, A. Heck and F. Murtagh, eds., European Southern Observatory, Munich, 319–324, 1992.

14. Heck, A., ed., *Desktop Publishing in Astronomy and Space Sciences*, World Scientific Publishing, Singapore, 1992.

15. Jenkner, H., "Database aspects of the Guide Star Catalog", in *Databases and On-line Data in Astronomy*, M.A. Albrecht and D. Egret, eds., Kluwer Academic Publishers, Dordrecht, 59–66, 1991.

16. Lamport, L., LaTeX *A document Preparation System*, Addison-Wesley, Reading, MA, 1986.

17. Pasian, F. and Richmond, A., "User interfaces in astronomy", in *Databases and On-line Data in Astronomy*, M.A. Albrecht and D. Egret, eds., Kluwer Academic Publishers, Dordrecht, 235–252, 1991.

18. Péron, M., Ochsenbein, F. and Grosbøl, P., "The ESO-MIDAS table file system", in *Astronomy from Large Databases II*, A. Heck and F. Murtagh, eds., European Southern Observatory, Munich, 433–438, 1992.

19. Pirenne, B., Albrecht, M.A., Durand, D. and Gaudet, S., "STARCAT: an interface to astronomical databases", in *Astronomy from Large Databases II*, A. Heck and F. Murtagh, eds., European Southern Observatory, Munich, 447–453, 1992.

20. Pirenne, B. and Albrecht, M.A., "The archive column", *ST-ECF Newsletter*, **19**, 1993.

21. Pirenne, B., "Two years of HST archive operations – first conclusions", *ST-ECF Newsletter*, **19**, 1993.

22. Raimond, E., "Archives of the Isaac Newton Group, La Palma and Westerbork observatories", in *Databases and On-line Data in Astronomy*, M.A. Albrecht and D. Egret, eds., Kluwer Academic Publishers, Dordrecht, 115–124, 1991.

23. Raimond, E., *private communication*, 1992.

24. Rey-Watson, J., "Astronomical bibliography from commercial databases", in *Databases and On-line Data in Astronomy*, M.A. Albrecht and D. Egret, eds., Kluwer Academic Publishers, Dordrecht, 199–210, 1991.

25. van Herwijnen, E., *Practical SGML*, Kluwer Academic Publishers, Dordrecht, 1990.

26. Wamsteker, W., "The many faces of the archive of the International Ultraviolet Explorer", in *Databases and On-line Data in Astronomy*, M.A. Albrecht and D. Egret, eds., Kluwer Academic Publishers, Dordrecht, 35–46, 1991.

27. Weiss, J.R. and Good, J.C., "The NASA Astrophysics Data System",in *Databases and On-line Data in Astronomy*, M.A. Albrecht and D. Egret, eds., Kluwer Academic Publishers, Dordrecht, 139–150, 1991.

28. White, N. and Giommi, P., "The EXOSAT database system", in *Databases and On-line Data in Astronomy*, M.A. Albrecht and D. Egret, eds., Kluwer Academic Publishers, Dordrecht, 11–16, 1991.

Chapter 12

Multistep Queries: The Need for a Correlation Environment

Mike Hapgood

Rutherford Appleton Laboratory
Chilton, Didcot
Oxfordshire, OX11 0QX (United Kingdom)
Email: cym@ibm-b.rutherford.ac.uk

12.1 Introduction

It is now widely recognised that there is a need for good methods of disseminating the large amounts of digital data generated by large scientific projects such as space missions and major ground-based observatories. Funding for these large projects can only be justified if there is a good scientific return on the data that they collect. Thus, wherever possible, the data from a project should be made widely available within the user community for that project (subject to any time-limited period of individual privilege).

One important aspect of improving scientific data dissemination is to develop better methods of data retrieval so that individual users can easily identify the dataset that they require and transfer it to their home institute for analysis and publication. This chapter examines the way in which many scientific data retrievals are carried out. It shows that this naturally leads to a multi-step query process in which queries are progressively refined. This, in turn, leads to a number of specialised requirements for scientific data retrieval. One of these is a requirement for a sufficient set of analysis tools to enable the user to examine the result of each query step and make a scientific judgement on how to formulate the next step. These tools should be integrated with each other, and with the data query system, so that it is easy to use existing tools and straightforward to add new tools. It is this integrated set of tools, together with any underlying software infrastructure, that we term the "correlation environment" or CE. Note that the term correlation is

A. Heck and F. Murtagh (eds.), Intelligent Information Retrieval: The Case of Astronomy and Related Space Sciences, 153–171.

used here in its broad scientific sense – a correlation is an apparent connection between observations of phenomena that suggests the existence of a law of nature linking those phenomena. The CE "toolkit" should not be restricted to the mathematical functions of autocorrelation and cross-correlation but would, of course, include those functions.

The concept of a correlation environment originated in studies for the European Space Information System or ESIS which is now being developed by ESA (Albrecht et al., 1988; Giaretta et al., 1991). However, the reader should note that this chapter does not describe the CE implemented for ESIS. Instead this chapter attempts to look, independently of ESIS, at some general issues arising from the CE concept: what should it do? what are its interfaces? how can it operate in a distributed environment? For a description of the particular CE implementation now being developed for inclusion in the ESIS pilot system, the reader is referred to the papers by Giommi et al. (1992) and Ciarlo et al. (1992).

Finally some caveats:

- The ideas put forward in this chapter are tentative. The author believes that they are good solutions to the problems of data retrieval but does not wish to exclude other possible solutions. Discussion of these ideas should be encouraged.
- This chapter is mainly concerned with retrieval operations on actual data, i.e. sequences of measurements of physical quantities, and not with catalogues of objects or events.
- All the examples are expressed in terms of space physics, i.e. the study of the regions of ionised matter in the solar corona, in interplanetary space, in planetary magnetospheres and ionospheres, and in comets. However, we note that the use of a CE in astronomy has been successfully developed (see, for example, Giommi et al., 1992).

12.2 What are Multi-Step Queries?

The concept of multi-step queries arises naturally when databases are used in scientific research. A researcher investigating a scientific problem will identify a dataset (or datasets) that may be used in his or her exploration of that problem. They will perform a query on a dataset, examine the results of that query and then make a scientific judgement about these results. In the early stages of exploration, that judgement is likely to be a new query or, perhaps, a refinement of the original query; a final result will, in general, be achieved only after stepping through many queries. If multiple datasets are used, some of these steps will involve merging datasets. The scientific result may be the final query itself and not the data which are extracted by it.

Thus, if we construct a model to represent how databases should be used in the exploratory phase of scientific research, that model should contain a feedback loop in which the results of one query are used as an input to the next query — to assist the specification of the procedures applied in the next step and/or to act as the data queried in that step. A graphical representation of this model is shown in Figure 12.1. It is this sequence of queries in the feedback loop that we refer to as multi-step queries. Note that this process of stepwise exploration is simply a restatement of the scientific method in the context of databases. It is a software equivalent of experimental physics, which we might term "experimental databasing".

The multi-step approach described above is very different from that used in commercial and administrative applications of databases, where the aim is to retrieve data as quickly as possible. Queries in these areas usually operate in a single step. For example, when checking in for an airline flight, you expect the check-in operator to retrieve the record relating to your journey by entering the flight number and your name. This can be done in a single step using a logical condition on these two parameters to search for and select the appropriate record. It is this type of query that the standard database query language (Structured Query Language or SQL) is primarily designed to handle. An implicit idea behind the standard is that queries go from the database to the result in a single step. To achieve this SQL allows the user to build very complex operations within a single query. If the query conditions have to be changed in the light of results, the whole query must be repeated using the new conditions.

The multi-step approach to scientific queries is therefore very different to the standard SQL approach to database queries. This leads to a number of special requirements. The key requirement that we consider here is the need to examine intermediate results. This task requires that a set of basic analysis tools is integrated with the query software. As indicated in the previous section, this set of tools is the "correlation environment" or CE.

Another important requirement that arises from the multi-step approach is the need to hold intermediate results for input into the next step. These may be held in a real database (i.e. in a separate file) or as a set of pointers that indicate which records of the initial database form the active subset. In some cases, the intermediate results may be held for a single step and overwritten once a new query is executed. However, in other cases, it may be desirable that intermediate results are held more permanently for later use. This permits a user to construct sophisticated sequences of queries as illustrated in Figure 12.2, e.g. branching queries in which several final results are obtained by different operations on an intermediate product; branching queries that are then joined again.

In the following sections we will discuss some of the tools that are needed in a correlation environment. We will also discuss the requirements for interfaces to a CE, in particular those between a CE and the main part of its query system – to transfer data to the CE and to transfer feedback to the query system in order to refine queries. Finally we will discuss the possible operation of a CE in a distributed environment.

12.3 Testing of Intermediate Results

The discussion in this section is expressed in terms of operations on ordered tables since these are the form of data with which the author is most familiar. Thus the discussion may need to be extended to cover more complex data formats. We define an ordered table to be a set of records each of which contains the same set of fields and one of these fields is an ordered key, e.g. the time tag in many space physics datasets. The fields in an ordered

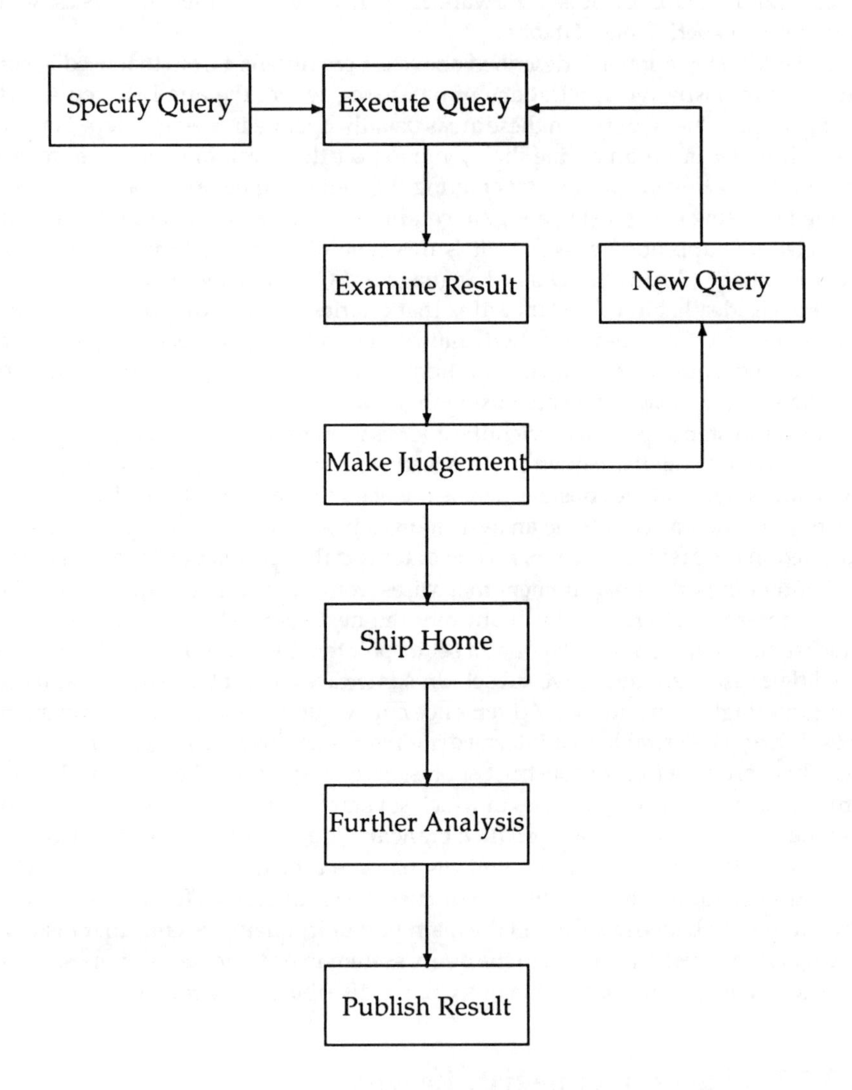

Figure 12.1: Model of scientific database queries showing the feedback loop for query refinement which leads to the multi-step approach.

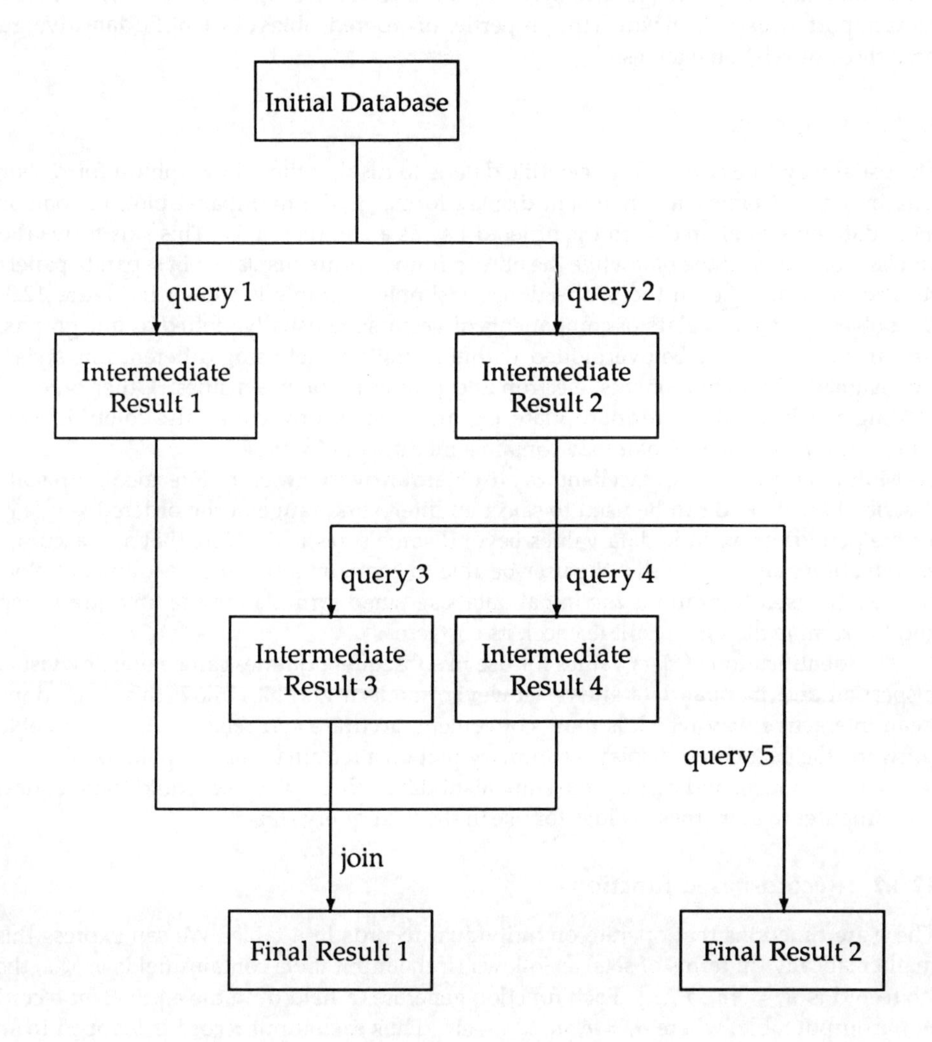

Figure 12.2: Models of branching and rejoining queries.

table, including the key, may be scalars, vectors, tensors, images or multi-dimensional arrays – in fact, any well-defined data object.

Ordered tables, as defined above, are very similar to the relational tables used in most modern database systems. Indeed, ordered tables may be conveniently held in a relational database management system. However, as we will show below, there are some important aspects in which the properties of ordered tables of scientific data diverge from those of relational tables.

12.3.1 Graphics

The usual way to examine any scientific data is to display them in graphical form. For data in ordered tables, a convenient display format is the multipanel plot, i.e. one or more data objects plotted with the ordered key as a common axis. This axis forms the abscissa, or x axis, of the plot while the other parameters are displayed in separate panels stacked vertically, i.e. in the y direction. A simple example is shown in Figure 12.3. Data objects such as scalars or components of vectors are usually plotted as line graphs; similar quantities may be overplotted within a single panel using different line styles, e.g. magnetic field components, electron and proton number densities. Other types of plotting may be used where appropriate, e.g. arrow plots for vectors, false-colour images for spectra. A multipanel plot may contain a mixture of plot types.

Multipanel plots are an excellent way to obtain an overview of the intermediate results described above and can be used to select an interesting range of the ordered key (e.g. a time period) or exclude data values beyond some threshold. Note that any accurate examination may require that the user be able to zoom on any panel. Multipanel plots can also be used to examine the initial database when formulating the first query step and to examine the final result to check its correctness.

The identification of data values for use in subsequent queries can be done by visual inspection and manual data entry. However, modern graphics allow this to be done in an interactive way which is more convenient, accurate and reliable. Given suitable software, the user could display a summary plot on a terminal, mark a point on the plot (e.g. with a mouse) and display the equivalent data values. The user could then request the computer to store these values for use in the next query step.

12.3.2 Record-based functions

These are functions that operate on individual records in a table. We can express this mathematically (in terms of sets) as follows. If the input table contains fields a, b, ..., the ith record is $R_i = \{a_i, b_i, ...\}$. Each function generates a field α_i in the equivalent record in the output table, where $\alpha_i = \alpha(a_i, b_i, ...)$ etc. Thus each input record is mapped to an output record; this output record is independent of any other input record.

Examples of such functions include: (i) mathematical manipulation to obtain derived parameters, and (ii) coordinate transformations. To give concrete examples:

1. Suppose that a user wishes to study the effects of solar wind pressure pulses on Earth's magnetosphere. To identify periods of interest, he or she must search

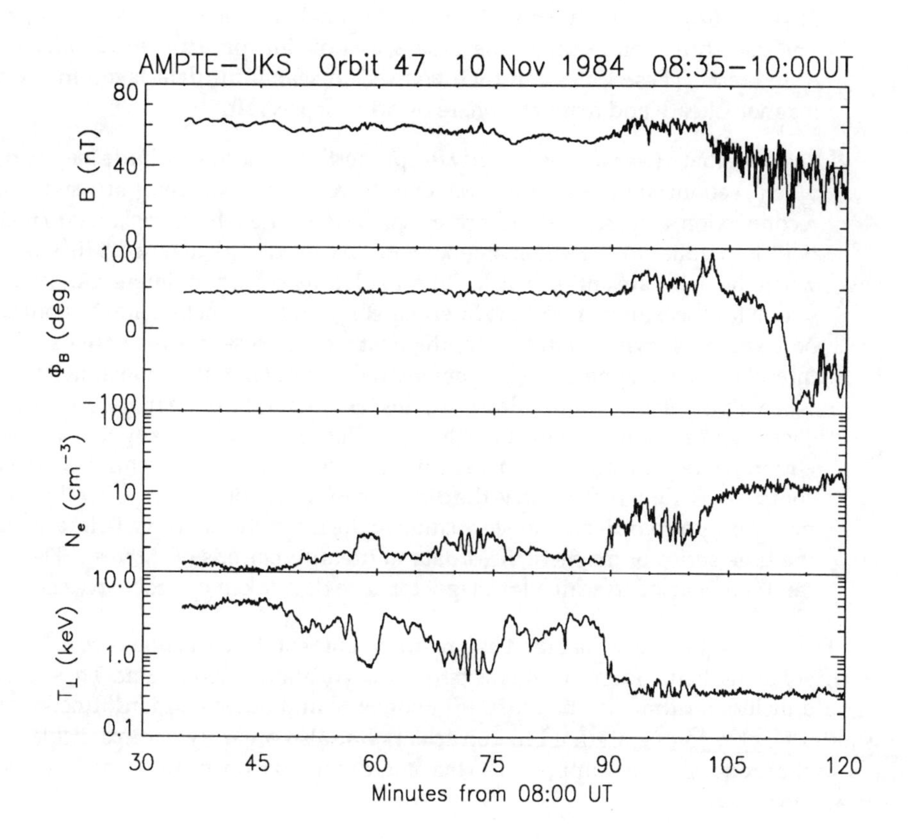

Figure 12.3: Example of a multipanel plot showing in-situ space plasma measurements made during an outbound crossing of the Earth's magnetopause by the AMPTE-UKS spacecraft. Starting at the top, the four panels show: (i) the total magnetic field strength; (ii) the azimuth, Φ_B, of the magnetic field in a plane tangential to the magnetopause ($\Phi_B = 0$ indicates that an azimuth co-planar with Earth's magnetic dipole axis); (iii) the electron number density; and (iv) the electron temperature perpendicular to the magnetic field. These data were obtained from the UK Geophysical Data Facility and plotted using IDL.

for measurements of the solar wind plasma made just outside the bow shock. Having found these, the next step is to calculate ram pressure from solar wind plasma data — ion number density n and bulk velocity v — using the formula $P_{\rm ram} = 1.67 \times 10^{-6} n v^2$ (where n is in $\rm cm^{-3}$, v is in $\rm km\,s^{-1}$ and $P_{\rm ram}$ is in nanopascals). The user can then examine the time series of ram pressure to identify periods of interest. These times can then be used to search for data taken inside in the magnetosheath and magnetosphere by other spacecraft.

2. Suppose that a user wishes to study the properties of the low-latitude magnetopause when reconnection rates are low. Our current understanding of magnetopause reconnection suggests that these rates will be low when the interplanetary magnetic field is parallel to the geomagnetic field just inside the magnetopause and high when the fields are anti-parallel. Thus, to identify periods of interest, the user must search for measurements of the interplanetary magnetic field made just outside the bow shock. Having found these, the next step is to search for periods when the interplanetary magnetic field is preferentially parallel to the geomagnetic field, i.e. when it points northward. However, this test should be done in the Solar Magnetic coordinate system for which north is parallel to the Earth's magnetic dipole axis. Unfortunately, magnetic field data are usually stored in Geocentric Solar Ecliptic coordinates, for which north is the direction of the ecliptic north pole. Thus the user must apply a coordinate transformation to the magnetic field data before examining the time series of magnetic field data to identify periods of interest. These times can then be used to search for magnetopause data taken by other spacecraft.

These examples show that there is a requirement that the user should be able to apply a variety of mathematical functions as part of a correlation environment. These functions should include arithmetic operators, trigonometric and other standard functions. They should be able to operate not only on scalars but also on arrays. In particular, matrix multiplication should be supported so that coordinate transformations can be carried out (Hapgood, 1991).

12.3.3 Column-based functions

These are functions that operate on the columns in a table. In mathematical terms the input records are $R_i = \{a_i, b_i, ...\}$ as before, where there are n records in the input table. The output of the functions is a set of values of the form α_i where $\alpha_i = \alpha(A, B, ...)$ and $A = \{a_1, a_2, ...a_i, ...a_n\}$, $B = \{b_1, b_2, ...b_i, ...b_n\}$, etc. These output values may be fields in a table with similar structure to the input table, fields in a very different table or a single set of values. The prime examples of such functions are time series analysis functions. To give concrete examples:

1. Suppose that a user wishes to search for periods when there were strong low frequency waves in the magnetic field. This can be done by retrieving magnetic field data for the whole period of interest and passing it through a suitable filter to cut out high-frequency fluctuations and to remove any DC shift. Periods of strong

wave activity can then be identified by examining a plot of filtered magnetic field data. The filter takes one or more magnetic field components plus the time-tag as input columns. The output is a table similar in structure to the input table with the time-tag and the filtered magnetic field components as columns.

2. Suppose that a user wishes to identify periods when there were strong radial flows in the magnetospheric boundary layer. To do this the user must transform the plasma flow vector into boundary normal coordinates. Thus the first step is to calculate the matrix for this transformation. This may be done by a minimum variance analysis of magnetic field data (Hapgood, 1991, p. 716). Thus the user must retrieve magnetic field and plasma flow data, apply a minimum variance routine to the magnetic field data to determine the transformation matrix [1], and then apply this to the ion data using a record-based function as already specified. Any periods of strong radial flows can then be identified by examining the N component of the flow vector. The minimum variance routine takes all three magnetic field components as input columns. The output is a set of three eigenvalues and three eigenvectors, from which the transformation matrix can be derived.

Another possible application of column-based functions is the automatic recognition of events in time series. To do this the user will have to apply an algorithm which looks for patterns in the time series of one or more parameters. This is simply a function that takes several columns as input and outputs a table whose columns are the time of the event and any attributes assigned to the event. We do not discuss this application in detail because the development of event recognition algorithms is still in its infancy. However, note that the provision of a CE would provide a software environment which could facilitate the development of such algorithms.

It is this requirement to support column-based functions that marks the divergence of the ordered tables described here from the relational tables which are used in many database systems. The relational model (Codd, 1990) requires that each record in a table is totally independent of the other records in the table. This requirement is a desirable feature for most commercial and administrative databases, which are used for transaction processing applications. However, it is an undesirable feature for many scientific applications. This can best be seen by considering one instance of an ordered table – a table of time-tagged measurements. In a well-designed scientific experiment, the records in this table will be statistically independent but should be physically related so that one can study phenomena by comparing successive data records. In fact, this is the aim of time-series analysis. The requirement that records are physically related is just a re-statement of the Nyquist (sampling) theorem. Note, however, that this requirement does not require us to violate the concepts of database normalisation that are a fundamental aspect of relational tables. Normalisation should be applied to scientific data tables at the level of statistical, not physical, independence.

[1] But note that the user will need to use some CE tools to check that the results of the minimum variance analysis determine the boundary normal coordinate system with sufficient accuracy.

12.3.4 Models

Thus far, we have discussed correlation environment functions in terms of exploring a scientific problem using data from the query system. It has been implicitly assumed that these data are measurements made by one or more instruments. However, in many explorations of scientific problems it may be impossible to obtain some essential information from the available measurements. In many such cases, this obstacle can be overcome by obtaining that information from a mathematical model or numerical simulation. We may illustrate this by looking at the two examples discussed in the subsection on record-based functions. In both examples there was a requirement for the user to search for measurements made outside the Earth's bow shock. If the query system contains a catalogue of observed bow shock crossings, this condition may be found by examining that catalogue. However, if there is no such catalogue, the user will have to use a model of the bow shock to estimate when crossings occurred and then confirm this by examining the data around the estimated crossing time.

Thus there is a requirement that the CE should be able to obtain information from models of the phenomena under study. This requirement may be satisfied in one or two ways. First, information generated by models and simulations may be held in the databases accessible by the query system, retrieved in the usual way and imported into the CE. This approach, which is best suited to the results from major simulations, requires no special functionality in the CE. The second method is to run the models from the CE and generate the information as needed. This approach is well suited to simple mathematical models, such as the polynomials used to model magnetospheric boundaries, since these require only modest computer resources. It does, however, have important implications for CE functionality. The CE must provide facilities to select the model, to set its parameters, to run it and pick up the output into a form that can be manipulated by other CE functions such as graphics.

12.4 Interfaces

The successful operation of a correlation environment requires that it have well-defined interfaces with the query system and the user. This section explores these interfaces. A graphical representation of the interfaces discussed below is given in Figure 12.4.

12.4.1 Interface with the query system

Data import from the query system

First and foremost, a correlation environment must be able to import data from the query system – for display and for manipulation. This should be quite straightforward – the interface should pass a pointer so that the CE can read the data directly. The interface must allow the transfer of metadata, i.e. the attributes that describe the data, as well as the data itself. Key attributes such as variable names, units, range limits are all required for use by CE functions.

Information for query refinement

A correlation environment must be able to feed information back to its query system so that refined queries can be executed. This interface is more difficult to specify since there is some choice over the level of complexity that it should support. For instance, should a correlation environment pass back: (a) whole queries ready for execution (expressed in the appropriate dialect of SQL); (b) logical conditions which are then incorporated into queries by the query system; or (c) data values which are then applied as conditions by the query system? In all cases the actual interface is quite simple. However, the different choices have implications for the complexity of the query system and CE software required to support the interface.

- If whole queries are passed back to the query system, the interface is to pass a command string. However, to generate this command string, the CE must replicate all the functionality for query formulation that is already supported in the query system.

- If logical conditions are passed back to the query system, the interface needs to pass a data structure containing three parameters: a variable name, a condition operator, and a data value. The query system side of the interface must include a facility to load the condition into the "current" query, where it may be the only condition applied or may be combined with other conditions via the usual Boolean algebra. To generate the condition the CE must allow the user to select data values, e.g. by clicking a mouse on a plot as described above, and to associate a condition operator with this value. Possible condition operators include: $=, >, <, \geq, \leq, \neq$, IN_RANGE. Note that for the IN_RANGE operator the "data value" is a pair of limit values. It is assumed that the variable name will be picked up automatically since this data attribute is included in the data imported into the CE.

  This approach could be extended by creating a logical condition that applies to a function of several variables. In this case the "variable name" passed to the query system would specify how the function is formed from atomic variable names in the original database. The execution of these more complex queries would require that the query system support the evaluation of these functions.

- If data values are passed back to the query system, the interface needs to pass a data structure containing only two parameters: a variable name and a data value. These can be generated by the CE quite simply as for logical conditions. However, to incorporate this structure in a query, the query system must provide a facility to process it into a logical condition and then include that condition in the query.

The solution recommended here is the passing of logical conditions. The passing of whole queries is rejected because it seems undesirable to replicate the query formulation system in the CE. The choice between passing logical conditions and passing data values is finer. The former is chosen because it is simpler at a logical level; the data structure passed by the interface is handled as an entity by the query system. In contrast, the latter requires that the query system should process that data structure into a new form. The use of the higher level interface makes for a well-defined distribution of functionality between the query system and the CE.

Note that this solution is not necessarily unique. Other solutions besides the three discussed here may well be possible. However, it is argued that the ideas put forward here represent a good practicable solution.

Data return to the query system

Another requirement that we should consider is for the transfer of CE data products back to the query system. For example, some of the temporary databases generated using CE data manipulation tools could be of continuing scientific interest. Thus it would be appropriate if they could be stored in a permanent database which is accessed via the query system. However, this requirement is not essential to the operation of a CE but is only a desirable addition to CE functionality. Thus the provision of an interface to serve this purpose should be given lower priority than the other interfaces described here.

We should also note that the provision of this interface could be quite demanding:

- The transfer of the new data products into permanent databases would require that these products are adequately described – in particular, in terms of database attributes but also in terms of the traceability of the data source(s). This would have a significant impact on CE functionality: (a) it would require that the data manipulation tools keep track of the necessary information; (b) it would require the provision of tools to prompt users for descriptive information to be attached to data products.

- A large amount of data and metadata would have to be transferred through the interface.

- The query system would require facilities to manage the incoming data products (authorization, cataloguing and archiving).

It would, therefore, be difficult to provide a general-purpose interface to handle data return to the query system. However, it may be possible to provide specialized interfaces to handle the return of specific data products of general interest. In such cases, the data returned could be required to conform to some predefined structure and be stored not in a separate new database but be appended to some existing database. In this case, the user could select which type of data they wish to return (perhaps from a menu); the CE would then select a "template" for the data structure to be returned, fill in all items available automatically and prompt the user for input where needed; on completion the data structure would be transferred through the interface and appended to the appropriate database. If required, the query system would check if that user was authorised to update the database.

A good example would be a catalogue of events e.g. magnetopause crossings. The fields in the data structure might include: spacecraft name, data and time of crossing, type of data used to identify the crossing, name of user making the identification, date and time of the identification. In this case, the user could examine a multipanel plot, pull down an "event" menu, select "magnetopause crossing" and then click a mouse at the point (time and data panel) on the display where the crossing has been identified. The software should then be able to pick up the spacecraft name, date and time of crossing and type of data from the plot. The name of user should be known from the account name and the date and time of the identification from the system clock.

Another possible way of returning data to the query system would be to place the returned data in a personal database in the user's own filespace. Such databases would not form part of the permanent archive and so would not be supported for general access. However, they would operate under the query system software so that the user could manipulate them as required.

12.4.2 User interface

User interaction

Any implementation of a correlation environment will require a good user interface. It must support the standard elements that are required to support user interaction with the software, namely:

- Command interface – to allow the user to control CE operation. This could be a simple command line interface, a menu system or a modern graphical interface (icon and mouse).
- Display interface – to display command prompts, CE status and results. This interface must support modern graphics including false colour images and a pointing device.

A highly desirable requirement on the user interface is that the user be able to switch easily between the CE and query system user interfaces. This is particularly important when refining queries. The user should be able to refine logical conditions with the CE and then switch systems to see the conditions incorporated in the current query. With the increasing use of window systems this requirement is easily satisfied. The user can open separate windows for the query system and CE user interfaces. Thus both interfaces may be viewed at once and it is trivial to switch between them.

The CE user interface may be provided either as a subset of the user interface for the main query system or as a completely separate system. However, this distinction is becoming less important with the advent of window systems. As indicated above one can open separate windows for the two systems and, from the user's point of view, it is irrelevant how the two windows are generated.

Data export

The CE user interface should also support data export to the user. This will allow the user to extract the results of CE operations and ship them to his or her home computer

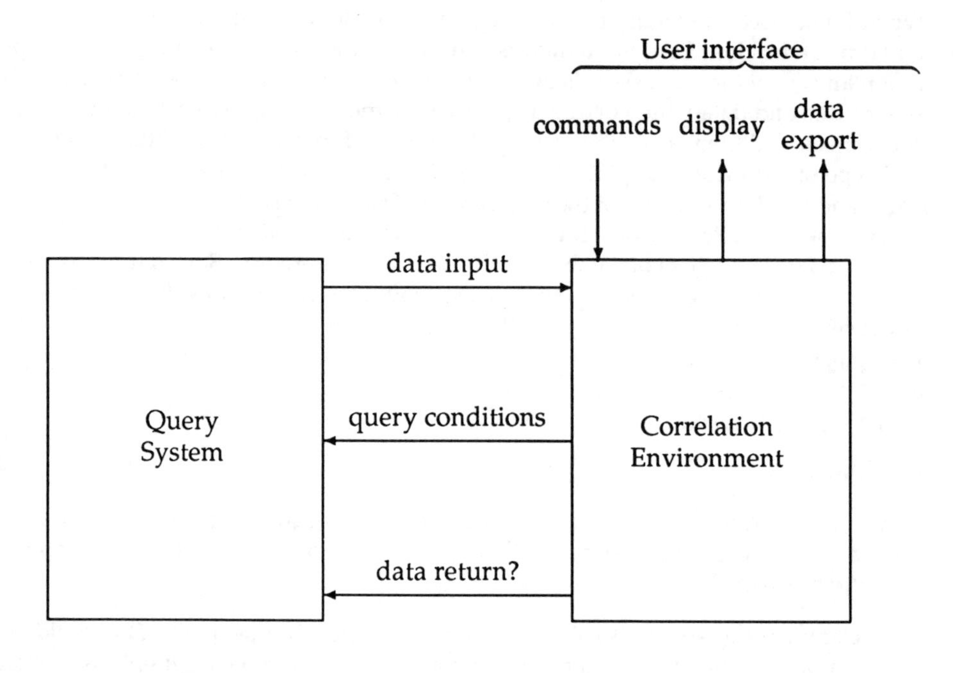

Figure 12.4: Correlation environment interfaces with query system and the user.

for further analysis and publication.

This interface may be implemented with varying degrees of sophistication depending on the degree to which descriptive information is attached to the exported data. The latter is a desirable but not essential requirement of data export. Thus this interface could be quite simple with data output in ASCII files without descriptive information. This approach would be appropriate to a minimal implementation of a correlation environment. A more sophisticated implementation might include descriptive information; this would require the exported data to be output in a standard format which supports descriptive information, e.g. NSSDC's Common Data Format (CDF; see NSSDC, 1991)

Data import

There may also be a requirement for data import into the CE. However, this might best be supported via a facility to import data into the query system and then transfer them to the CE. This route ensures that the full power of the query system and the CE can be applied to the data.

12.5 Operation in a distributed environment

When used in a distributed data system, the correlation environment software should be designed to make efficient use of the intervening network. In particular, it should minimise network traffic. In this section we examine the implications of this requirement for each of the types of tools discussed in section 12.3.

12.5.1 Graphics

When generating plots from a remote database, one should consider whether it is best carried out by moving the database to the user's computer and generating the plot there, or by generating the plot remotely and shipping it over the network in a compact format such as a computer graphics metafile (CGM). We expect that, in general, the latter should put less load on the network since: (a) the database may contain more fields than the plot; (b) the summary plot will usually have lower resolution (in both x and y) than the data. Thus we have assumed that this is the preferred strategy. However, we note that this argument only applies if the graphics file is compact; this is not true of all formats, e.g. Postscript files are relatively very large.

To implement this function in a distributed system the user's local software should allow them to specify plot parameters, e.g. for a multipanel plot these are the names of the parameters to be displayed and the range of the ordered key. This information should then be dispatched to the remote node where the plot is constructed and output into a suitable file, which is returned to the local node for display. The file should contain not only the basic plot but also information relating plot coordinates to data values. This is required to support selection of data values by a pointing device. The file should distinguish the separate panels so that these can be zoomed separately.

12.5.2 Record-based functions

The implementation of these functions in a distributed environment should be quite straightforward. It is clearly more efficient to keep the data at the remote node; the network need only carry the small amount of information specifying the functions to be applied. The user's local node should support the formulation of these functions, which can then be transmitted over the network for execution. The result is a new temporary database which can be examined using graphics tools.

The software to support the formulation of mathematical functions may be quite sophisticated. It should provide facilities to specify coordinate transformations and to allow the user to check these before transmission to the remote node. It may also support pre-defined functions, e.g. the calculation of standard plasma quantities such as Debye length, gyroradius, Alfvén speed, etc. The procedure for sending function specifications to the remote node should eliminate redundant information, thereby minimising network traffic. For example, the nine terms in a coordinate transformation matrix can be specified by only three independent parameters.

12.5.3 Column-based functions

The implementation of these functions is similar to that of the record-based functions discussed above. Again it is clearly more efficient to keep the data at the remote node and transfer only the small amount of information specifying the functions to be applied. Thus the user's local node should also support the formulation of these functions and their despatch for execution at the remote node. The main difference from record-based functions is that the result may be: (a) a small set of values which should be transferred to the user for visual examination; (b) a new temporary database which can be examined using graphics tools; and (c) both of the previous outputs.

12.5.4 Models

The distribution of functions to generate data from models will depend on the quantity of data generated by each model and on the application of those data. If these data are used independently of other datasets (e.g. to help the user estimate the time of a bow shock crossing), the models would best be run on the user's local computer. However, if the data are to be merged with other datasets, the models would best be run on the computer holding the data to be merged. In this case, the CE on the local computer must allow the user to select the model, choose its parameters and send all the necessary information to the remote node for execution.

12.5.5 General requirements

The discussion above suggests the following general requirements for the operation of a correlation environment in a distributed environment:

1. Each intermediate result, as specified in section 12.2, should be held on the remote computer where it was generated. Intermediate results should be transferred over the network only: (a) when explicitly requested by the user; and (b) when they have to be joined with datasets located at other nodes.

2. To test these intermediate results, the CE software tools must execute on the remote computer. This software must be capable of receiving commands from the user's computer.

3. The results of these tests should be transferred over the network in as compact a form as possible. The CE software on the user's computer must be able to receive and display these results.

These requirements indicate that CE functionality should be distributed over the system with command and display software running on the user's computer and a copy of the CE tools operating on every node that can hold data.

12.6 Summary and Conclusions

This chapter has shown that the ways in which databases may be used in the exploration of scientific problems can be very different to the ways in which databases are used in commercial and administrative applications. Scientific exploration is likely to require a sequence of queries which mirrors the scientific method in experimental science, namely that the user examines the results of each query and then uses his or her scientific judgement to formulate the next query. To support these "multi-step queries" the database must include a sufficient set of analysis tools to allow the user to examine results and look for "correlations" between phenomena. For ease of use and development, these tools must be properly integrated with each other and with the underlying database system(s). Hence we term this set of tools a "correlation environment" or CE.

It is shown that a CE should contain a wide range of tools including graphics, data manipulation functions and access to basic models of physical phenomena. The data manipulation functions should include not only functions that operate on individual records within a data table but also functions that operate on columns, e.g. time series analysis on time-tagged records. The use of the latter functions is not consistent with the independence of records required in relational tables and indicates that the relational model cannot support all the functionality required in scientific databases. Modularity of CE functions will be important to ensure that new tools can be added as and when appropriate. These modules should access the CE infrastructure via application programming interfaces so that new modules can be added by skilled users as well as by system developers.

To support this functionality a CE must have well-defined interfaces with the query system and the user. The latter would be a standard user interface supporting command input, screen displays and data export as outputs; it is recommended that data import be carried out only via the query system. The interface with the query system is more specialised since this must support data input into the CE, the export of new query conditions to the query system and, possibly, the export of selected data back to the query system.

It is worth considering briefly the possible future development of correlation environment concepts over the next decade. The reason for doing this is not merely to speculate but to help define the present objectives of a CE. However, speculation about software developments is fraught with examples of false predictions and so it is wise to make a caveat that the objectives of a CE must be reviewed periodically.

In the future we may reasonably expect much increased use of packages such as Matlab, IDL etc. and hence a requirement to be able to call these from a CE. These software packages are already modular and have data import and export facilities which are much easier to use than was the case say, a decade ago. However, they are not generally oriented towards database use, e.g. the importation of ASCII tables into IDL can be cumbersome (IDL, 1992). However, the development of better database interfaces is perhaps only a matter of time, e.g. NSSDC will soon release an interface between IDL and its CDF data storage format (Groucher, 1992). These packages already have many of the CE facilities but the essential function which they do not include is the multi-step query formulation which has been stressed in this chapter. The links to such packages must emphasize the need for a modular CE for all the usual software engineering reasons. One should aim to make it possible to switch between packages at least as easily as we can now transfer documents between different modern wordprocessing packages. All of this is involved in the use of the word "Environment" in the term CE.

What of "Correlation"? As discussed above, it must be taken to include all methods by which scientists establish connections between phenomena and not just the mathematical concepts of auto- and cross-correlation. Existing methods such as graphics will naturally develop during the next decade. CE toolkits should be upgraded regularly to take advantage of these developments. It also seems likely that many new methods for exploring scientific data will become widely available during the next decade. These include mathematical approaches such as Narmax which are proving effective in many fields of identification (Billings, 1986; Billings and Chen, 1989) and the various "biological" algorithms such as neural networks, genetic algorithms and fuzzy logic. These techniques should be included in CE toolkits as and when possible. We should note that modularity of CE software is the key factor in facilitating such developments.

Finally, it must be stressed that a CE should not attempt to become a universal tool to solve all scientific problems! It should concentrate on key functionality, have good modularity and interface to other packages.

12.7 Acknowledgements

My particular thanks to Les Woolliscroft who made many helpful and stimulating comments on the draft of this chapter. My thanks also to Miguel Albrecht who first introduced me with enthusiasm to the idea of a correlation environment when developing the initial requirements for the European Space Information System (ESIS) and to David Giaretta whose thinking sharpened many of the concepts of a CE.

References

1. Albrecht, M., Russo, G., Richmond, A. and Hapgood M., *Towards a European Space Information System: Volume 1, General Overview of User Requirements*, 21, ESA-ESRIN, Frascati, Italy, 1988.

2. Billings, S.A., "Introduction to nonlinear systems analysis and design", in *Signal Processing for Control*, Godfrey, K. and Jones, P., eds., 263–294, Springer-Verlag, Berlin, 1986.

3. Billings, S.A. and Chen, S., "Extended model set, global data and threshold model identification of severely nonlinear systems", *International Journal of Control*, **50**, 1897–1923, 1989.

4. Ciarlo, A., Ansari, S., Donzelli, P., Giommi, P., Regener, P., Stokke, H., Torrente, P. and Walker S., "The European Space Information System – approaching pilot-

project evaluation", *ESA Bulletin*, No. **72**, 99–106, 1992.

5. Codd, E.F., *The Relational Model for Database Management, Volume 2*, Addison-Wesley, New York, 1990.

6. Giaretta, D.L., Hapgood, M.A., Lepine, D.R., Penny, A.J., Read, B.J., Alleyne, H.St.C., Walker, S.N. and Woolliscroft, L.J.C., *European Space Information System: Correlation Environment*. Report on ESA contract RFQ/3-6600/89/HGEI, Rutherford Appleton Laboratory and University of Sheffield, 1991.

7. Giommi, P., Ansari, S.G., Ciarlo, A., Donzelli, P., Stokke, H., Torrente, P. and Walker S.N., "The ESIS correlation environment prototype", in *Astronomical Data Analysis Software and Systems I*, Worrall, D.M., Biemesderfer, C. and Barnes, J., eds., Astronomical Society Pacific, San Francisco, 59–61, 1992.

8. Groucher, G., "CDF News: Version 2.3 is ready for release", *NSSDC News*, **8**, 8, 1992.

9. Hapgood, M.A., "Space physics coordinate transformations: a user guide", *Planetary and Space Science*, **40**, 711–717, 1991.

10. *IDL User's Guide for IDL version 2.2*, edition of April 1, 1992. pp. 13-29 to 13-32, Research Systems Inc, Boulder, Colorado, 1992.

11. *NSSDC CDF User's Guide for VMS Systems, version 2.1*, NSSDC/WDC-A-R&S Report, **91-31**, NSSDC, Greenbelt, MD, 1991.

Chapter 13

Intelligent Information Retrieval in High Energy Physics

Eric van Herwijnen

CERN
CH-1211 Genève 23 (Switzerland)
Email: eric@cernvm.cern.ch

Abstract

The number of articles published by physicists is steadily increasing, partly due to the "publish or perish" principle. The number of printed documents is also growing, despite the claims of computer science that the paperless office is around the corner. In this chapter I discuss some of the problems in the electronic publishing area and review tools and technological developments which are currently available to help a scientist retrieve information.

13.1 Introduction

The exchange of information within a (the) scientific community, with the publisher as intermediary, is shown in figure 13.1. The authors of research papers are the providers, P. The publishers are the gatherers of information, G. They accept information from many providers, gather it in the form of a journal issue, and distribute it. In this process, the publisher provides a quality check via the system of peer reviewing, a copy-editor assures notation is consistent, and in some cases improves the prose. The information is distributed to a group of consumers, C, where set C is a superset of set P.

Although this system has been accepted by the scientific community for about a hundred years as an essential part of humanity's effort to increase knowledge, there are now some signs that it is no longer functioning optimally.

A. Heck and F. Murtagh (eds.), Intelligent Information Retrieval: The Case of Astronomy and Related Space Sciences, 173–191.

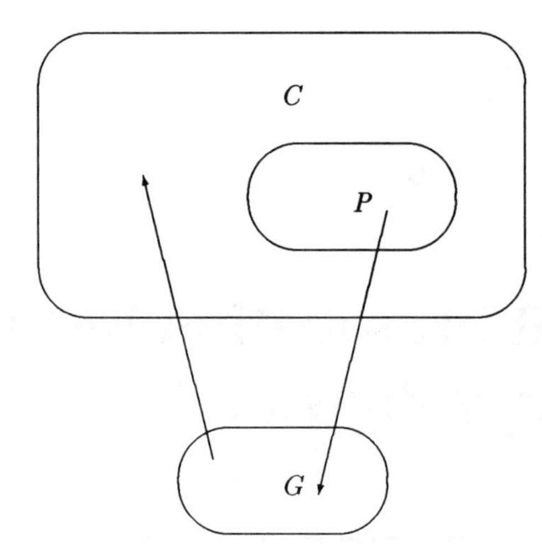

Figure 13.1: Information exchange within a (the) scientific community.

Some problems associated with the publishing process, i.e. the arrow pointing from $P \to G$ are:

1. The time between submission of the manuscript and its publication in a printed journal can be unreasonably long.
2. The variety of computer systems used by authors slows the publication process down rather than speeding it up.
3. The peer review process is undemocratic and unsuited for detecting fraud.

The exchange of information between authors and publishers is being streamlined by standardizing on common, electronic formats (see section 13.2). One may, therefore, hope that publishing will become easier and faster in a not too distant future.

An immediate solution to the slowness of the publication process in the high-energy physics world is the *preprint* system (see Figure 13.2).

When an article is submitted for publication to a journal, an unrefereed version called a *preprint* is distributed by the providers, P, to a group of consumers, C, usually colleagues working in the same field. Some physicists claim preprints are more important than published articles. This is supported by the fact that 75% of all reading-time in CERN's central library is spent on preprints. The preprint system, however, complicates the information exchange cycle and causes additional problems.

Some problems associated with the distribution of knowledge, i.e. the arrow pointing from $G \to C$, are:

1. The number of articles is growing at an enormous rate, pushed by the "publish or perish" principle.

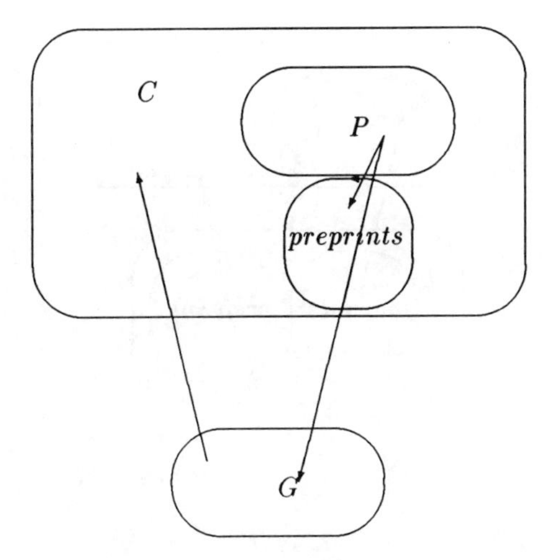

Figure 13.2: Information exchange within the high energy physics community.

2. The high number of contributions makes the retrieval of information difficult.
3. The number of authors on certain papers written by large experimental collaborations can be large (several hundreds). The contradictory situation arises where an author has not read his own paper.
4. The length of the publication list as a measure of someone's scientific importance is meaningless in many cases.
5. The distribution of preprints is undemocratic and costly (printing and mail). They get lost or stolen from library shelves.
6. The price of journal subscriptions is high and increasing.

Due to the price of journal subscriptions and other practical reasons, research institutes use their *library*, L, to distribute information to their consumers, C. The position of the library in the information exchange cycle is shown in Figure 13.3.

Preprints cause libraries the following problems:

1. The material needs to be selected before it can be added to the library's collection. CERN, for example, receives much material which is unrelated to high energy physics.
2. Archiving preprints uses valuable floor space.
3. Adding preprints to a library database is costly (about 3 people to keep up to date with 1000 document entries/month).
4. Accented characters and mathematical symbols are difficult to put into a bibliographic record.

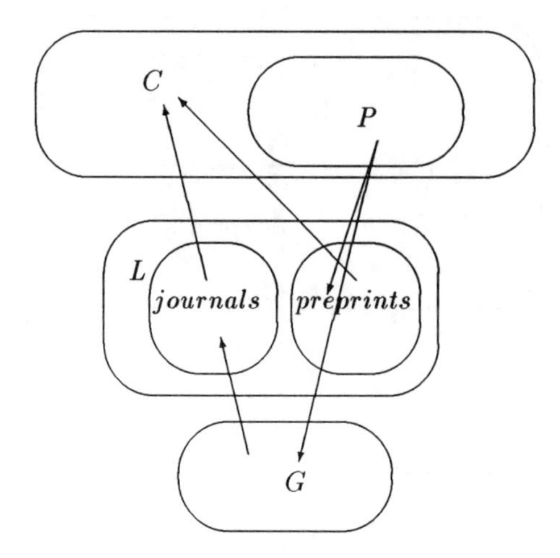

Figure 13.3: The role of the library for information exchange.

5. Once a preprint is published, it ceases to be a preprint (thus becoming an *anti-preprint*), and should be deleted from the database. There is no systematic way of finding out if and where a preprint is published.

6. Obtaining citations and analyzing them is an expensive exercise.

In what follows I discuss the problems which are caused by the interaction and overlap between article publication, preprint creation and distribution, and library use and management.

13.2 Article Publication

13.2.1 The need to move away from paper

The amount of published information is increasing rapidly. At present, about one million articles are published each year in the sciences and humanities. This number doubles every 15 years, and is therefore given by an expansion rate of $2^{t/T}$ where $T = 15$ and t is the number of years since the first article appeared in 1690 (Kircz, 1992). Assuming that on average an article has 10 pages, that it appears in a journal which has 1000 subscribers and that each journal is photocopied (recto-verso) 10 times, the total paper consumption is $10^5 \times 2^{t/16}$ A4 sheets (or $0.5 \times 2^{t/16}$ tonnes of paper) per year. The world's total paper production is 57×10^6 tonnes, and thus, if nothing happens, in about 180 years all the paper in the world will be used for the distribution of articles. It is therefore important that new media are explored as alternatives to paper for the distribution of information.

Although this pattern varies from discipline to discipline, a similar curve has been observed in high energy physics. At SLAC, for example, the number of preprints received by the library doubled between 1975 and 1992 (SLAC, 1992). It is imperative that new information retrieval techniques are introduced to cope with this un-digestible amount of material.

13.2.2 The need to standardize

In the move away from paper, a recent trend is that authors prepare their articles with text-processing software on some computer. This enables them to send the paper in electronic form (electronic manuscript or "compuscript") to the publisher. Publishers, however, are confronted with a variety of text-processing software on a variety of computer systems (van Herwijnen and Sens, 1989; Heck, 1992). Moreover, every scientific field appears to have its own "Top Ten" of most used text processing packages.

The conversion between these different systems is a costly effort, and the result is often that publishers still re-key manuscripts. The solution to this problem is for authors and publishers to exchange information in a single, universally available format. ISO, the International Organization for Standardization has proposed a standard for text exchange called SGML (van Herwijnen, 1990).

Before SGML can be used in this context, two requirements need to be met:

1. The structure of scientific articles has to be agreed on between authors and publishers in the form of an SGML Document Type Definition (DTD). The European Physical Society have selected for this purpose the DTD from the Association of American Publishers (ANSI/NISO, 1987). This DTD will become an ISO standard (ISO/DIS, 1992).

2. Vendors and developers of the most common word processors should support this DTD so that authors can use the system they are familiar with (such as TEX/LATEX, MS Word, WordPerfect, Framemaker, etc.) and obtain the SGML format without additional effort.

If the physics community decides to take this direction, a transition path should be defined which will enable authors to continue using their favourite text processor in such a way that publishers and libraries can convert documents to SGML.

13.2.3 Advantages of a standard storage format

If the communities of information providers, C, and gatherers, G, can agree on using a common exchange format such as SGML, both groups will see the following advantages:

- the article does not have to be retyped by the publisher, which saves time;
- the publisher can automatically extract data for its databases;
- the article can be resubmitted to other journals without the need for reformatting;
- by marking up the logical parts of articles, one can use new information retrieval techniques such as hypertext;

- libraries will be able to create full-text article databases (see section 13.3);
- when publishers make published articles available in SGML, libraries will be able to automate anti-preprinting.

13.2.4 Problems and limitations of standardization

When automating the publication process, several areas require special attention, irrespective of whether SGML is used as interchange standard.

1. Mathematical formulae are well handled by TEX, but not so well by other systems. There is a point to be made for an (at least partially) generic description of mathematics (Poppelier et al., 1992), since this enables searching of articles related to particular calculations. Perhaps one can also use this notation as input to symbolic algebra programs.
2. Pictures are handled more and more often by PostScript, but their inclusion in TEX and other documents is non standard and can cause problems.
3. In the peer review cycle, a guarantee is required that the document which is finally sent to the publisher, is identical to the document that was accepted by the referee. At least one authentification service, offered by Bellcore, exists (Charles, 1992), but this needs to be organized (see section 13.4.6).

The committee for publications of the European Physical Society is very active in this area, but we seem still to be several years away from a perfectly streamlined electronic publication process using SGML.

In the meantime, a growing number of publishers accept articles in TEX or LATEX. REVTEX from the American Physical Society (REVTEX, 1991) is a well known example; the American Astronomical Society are proposing AASTeX (Biemesderfer, 1992), and declare that SGML will be used for the production and storage of documents.

Publishers have started to recommend LATEX to authors with journal- or conference-specific style sheets. If the style sheet is correctly made, LATEX can be converted to SGML, and this could be a transition path to using SGML as an exchange format. There are some reasons why this approach is taking longer to catch on than expected:

- It is hard to persuade physicists to use LATEX with a style sheet if they already have a solution they are happy with, say PHYZZX (Weinstein, 1984). Physicists just want to automate and get their article out on paper in the way they like it. Forcing a format by introducing a stylesheet which they can't modify is not popular.
- The amount of support required for a stylesheet is often underestimated.
- To convert to SGML, a separate program is required for each journal's style sheet.

Although LATEX based solutions do not offer all the advantages of SGML, in the short term they are a step forward because they make people used to the idea of bypassing paper.

13.3 Automated Library Systems

Computers were first applied in libraries to catalogues and journal abstracts. Due to insufficient computer power, it took a while before satisfactory systems were developed.

Fairly complete systems now exist, such as the one used by CERN (Aleph, 1987) or SLAC (based on Spires, 1993). In addition to having catalogue information online, automated library systems are able to organize loans, recalls, and do accounting. The inclusion of periodicals (in addition to books) is complicated but can be done. For most libraries, the unit which is included is a journal issue. Information about individual articles and abstracts can be found by using services such as Inspec.

13.3.1 Shortcomings of automated library systems

Despite the increased use of computers in all areas of society, particularly those related to the production of (printed) text, the automation of libraries is still far from ideal. Some of the problems are:

1. There is no perfect system. In practice, users with similar interests develop their own solution (Spires). Commercial systems can be used, but often have to be adapted to local requirements (Aleph at CERN). Many other systems exist commercially and in the academic world.

2. Although they are quite complete from the librarian's point of view, they have complicated, sometimes obsolete, user interfaces. Client-server architecture-based systems with graphical user interfaces, that integrate well with networked PCs and Macintoshes, are rare.

3. Library systems are designed for university or public libraries, which have large collections of books and do many similar transactions (loans). These libraries are staffed. The CERN library has a small collection of books, a large collection of preprints, and is open 24 hours a day with staff present only during office hours.

4. It is difficult and time-consuming to integrate preprints. There is no satisfactory system to automate the treatment of anti-preprints.

5. It is difficult to integrate journals. Access to external databases exists, but supporting them is non-trivial.

6. Today's library systems are unable to cope with mathematical formulae, special characters, full-text documents and hypertext.

7. In future, books will be published electronically on CD-ROM, or otherwise. They should be integrated into library systems.

13.3.2 Advantages of SGML for library systems

By standardizing on SGML as an exchange format for preprints, libraries can piggy-back onto a system which gives advantages to both authors and publishers. SGML enables libraries to automatically extract the bibliographic information they require (van Herwijnen et al., 1992). In this way, significant savings in manpower can be achieved. To manually enter the 12,000 preprints into the library's database, 3 full-time staff were required at CERN in 1992. Since the number of preprints is increasing, this procedure must be automated. In addition, SGML solves the problem of special symbols in bibliographic data.

Anti-preprinting is an activity which is currently impossible to automate. For its annual report, CERN needs to know every year the complete collection of articles that were published by its staff and members of institutes participating in the CERN programmes. The only way to do this is to scan the weekly literature visually and investigate the origin of each article with CERN authors. The system fails if a preprint is published in conference proceedings or journals, which the library does not receive. For non-CERN publications, CERN relies on information gathered by DESY and SLAC (also manually and visually).

If published articles were available in SGML, the original reference number of the preprint would still be present in the source of the published article (although it would not appear in print), thus enabling anti-preprinting to be automated.

In addition to these managerial advantages (which are shared by libraries worldwide), SGML opens the door to a full-text hypertext database (see section 13.4.4). A document in SGML can be easily converted into a hypertext system with tools such as *Dyna*Text (1992).

13.4 Preprint Creation and Distribution

The many different systems which are used to create articles have been reviewed elsewhere (van Herwijnen and Sens, 1989; Heck, 1992). TEX based systems seem to be the most popular, although there are strong voices speaking in favour of user-friendly WYSIWYG systems which are available to people on their desk-tops.

With or without TEX, authors would benefit from having bibliographic information available on their desk-top to make it easier to make citations while writing articles. It is currently a cumbersome procedure to find the exact publication data of an article (or preprint). The BIBTEX system (which is a companion program to LATEX) is convenient, but not much used. If the complete literature were available as a BIBTEX file, this might change things.

Maintaining, modifying and using TEX is the kind of work that physicists consider fun, and most institutes have no difficulty in finding someone who is (or wants to become) a TEXpert. It is in the public domain and macro packages exist to do almost anything. The problems found by publishers because of lack of a standard exchange format, are not shared by the authors, who are quite happy with TEX because it is universally available.

It is difficult to predict if TEX will survive, or if not, what the future system will be, but it should be noted that TEX was not written as a WYSIWYG system. WYSIWYG editors which use TEX as a formatter have severe performance problems. On the other hand, the high energy physics world has been notoriously slow in catching up with other modern developments in computer science such as C++, claiming that past investment in Fortran is important to hold on to. It is likely that TEX will be around for many years.

13.4.1 Bulletin boards

In the 1980s academic networks, such as BITNET, EARN and Arpanet, linked the users of mainframe computers over wide areas. More recently, personal computers and workstations began to be used as front-ends and they were linked together as well as to the mainframes via Ethernet, Token Ring or Appletalk.

Documentation systems exploit new possibilities for information exchange. Examples on IBM VM systems are VM Tools, Grand and LISTSERV (Thomas, 1986). Documents are stored in central places and the user interacts via electronic mail to retrieve the required information. These systems have sophisticated distribution and notification mechanisms, enabling a group of users to share documents and posted comments. There are at present over 600 discussion groups on the network, and many of them are run by LISTSERV.

In parallel, Unix based systems have become popular in the academic world, in particular news systems on Usenet. They use the Internet network and TCP/IP communication protocols. These protocols allow remote logon with telnet and file access via ftp. [1] Many other types of *"bulletin-boards"* exist (Girvan, 1992).

In the field of Theoretical High Energy Physics, P. Ginsparg started a bulletin board in 1991 at Los Alamos, which automatically classifies new articles and notifies subscribers. Articles are stored in TEX, which means that a retrieved article needs to be formatted locally before it can be viewed or printed. It is not an accident that high energy physicists were pioneers in this area, since the enormous quantities of data they manipulate require a large degree of automatisation.

The advantage of document servers is that the distribution of a preprint is instantaneous and widespread. Moreover, the burden of printing falls on the reader instead of the author or the publisher. Considerable savings (in time, postage and paper) can be made by "printing on demand", assuming readers print only articles they want to read. Up to now, it was common practise for researchers to make their own photocopy of each preprint which they could lay their hands on, "just in case it got lost, stolen or famous".

To quote Ginsparg (1992), "...The system in its present form was not intended to replace journals, but only to organize a haphazard and unequal distribution of electronic preprints. It is increasingly being used as an electronic journal, however, given that it is frequently more rapid to retrieve electronically a file of a paper than to retrieve physically a paper from a filing cabinet."

[1]FTP is the Internet standard File Transfer Protocol. "Anonymous" ftp is a convention for allowing Internet users to transfer files to and from machines on which they do not have accounts, for example to support distribution of public domain software and documents. For security reasons users have restricted access to certain directories.

13.4.2 The CERN preprint server

There are now many physics bulletin boards, for various fields of physics. CERN has a document server which intends to be a complete database of high energy physics preprints in viewable and printable form. It is based on a Unix ftp/mail server with some added features. I shall describe some details of this prototype, because it shows how modern technology can help to solve some of the problems mentioned in the previous sections.

The database sits on the "Application Software Installation Server" server (a SUN Sparcstation 2 running SUN OS) at CERN. The primary function of this computer is the distribution of application software such as the CERN program library, TEX and other public domain software. The Internet address of this machine is `asis01.cern.ch`.

The files in the database originate from the following sources:

1. Official CERN preprints, which arrive at the printshop in PostScript form for printing.

2. All files from the Los Alamos servers (CERN shares the load of this with SLAC). We keep the TEX files and the PostScript files they generate.

3. Various ad hoc sources, e.g. whatever is received on the electronic mail address `PREPRINT@CERVM.CERN.CH` and can be converted to PostScript.

In 1992, we obtained 3,000 titles electronically, out of the total of 12,000 preprints received on paper by the library.

The files are stored in the directory `/preprints`. Subdirectories exist corresponding to each high energy physics institute, e.g. `/preprints/cern`. Each institute's subdirectory has a subdirectory `/ps` for PostScript files, e.g.
`/preprints/cern/ps/th.6422-92.ps`.

The filenames are identical to the reference numbers that appear on each preprint. Directories for other formats such as TEX, and SGML also exist.

13.4.3 Access to the server

The information in the CERN server can be accessed in various ways:

1. Anonymous ftp. Everyone connected to the Internet may ftp to the `asis01` server using the following commands:

   ```
   ftp asis01.cern.ch [128.141.201.136]
   userid: anonymous
   pwd: your_email_address
   cd preprints
   mget to obtain the files
   ```

2. Electronic mail. People without ftp access to `asis01` may obtain the files from the server by sending an electronic mail. For help, send an electronic mail message with the word "help" in the subject field to `preprint@asis01.cern.ch`

3. The preprints utility. This utility is the part that offers the added value. It allows interactive retrieval, online viewing and on demand printing of the articles, across wide area networks. It has a client-server architecture. The preprints utility can be used on any computer with an X Windows system (including PCs or Macs with X terminal emulators), connected to the Internet. At CERN, everyone connected to Ethernet has access. The following instructions start the system:

   The initial panel is shown in the figure below, to the right.

```
xhost +
telnet asis01.cern.ch
userid: preprint
pwd: hyper
preprints your-ip-number:0
```

13.4.4 Software requirements

`preprints` is written in TCL (Oosterhout, 1990), a light-weight interpretive command language. It uses Ghostview (Theisen, 1992), a front end to the Ghostscript PostScript previewer, both available in the public domain. Since the graphics is transmitted by the X11 protocol, the client-server architecture only requires the X11 software to be running in the client's machine.

`preprints` allows viewing and printing of individual files. The user can navigate through the directory containing files that belong to a given institute, preview a document, and, if desired, print it.

The CERN library's weekly preprint list contains an entry for each preprint received (on paper) by the library during the previous week, and is used as an index for the database. Working in close contact with the library ensures that all files on the server have an entry in one of the weekly lists.

The list is generated from Aleph in SGML form. *Dyna*Text provides a hypertext structure with links to PostScript files on the server. This "intelligent" version of the weekly list allows the user to retrieve any one of the preprints stored in the server.

*Dyna*Text presents two windows: a table of contents and a full text view of the document. By pressing on a section heading in the table of contents window, the full text view immediately jumps to that part. A query language permits full text keyword searching combined with queries about the document's structure such as titles, authors and reference numbers.

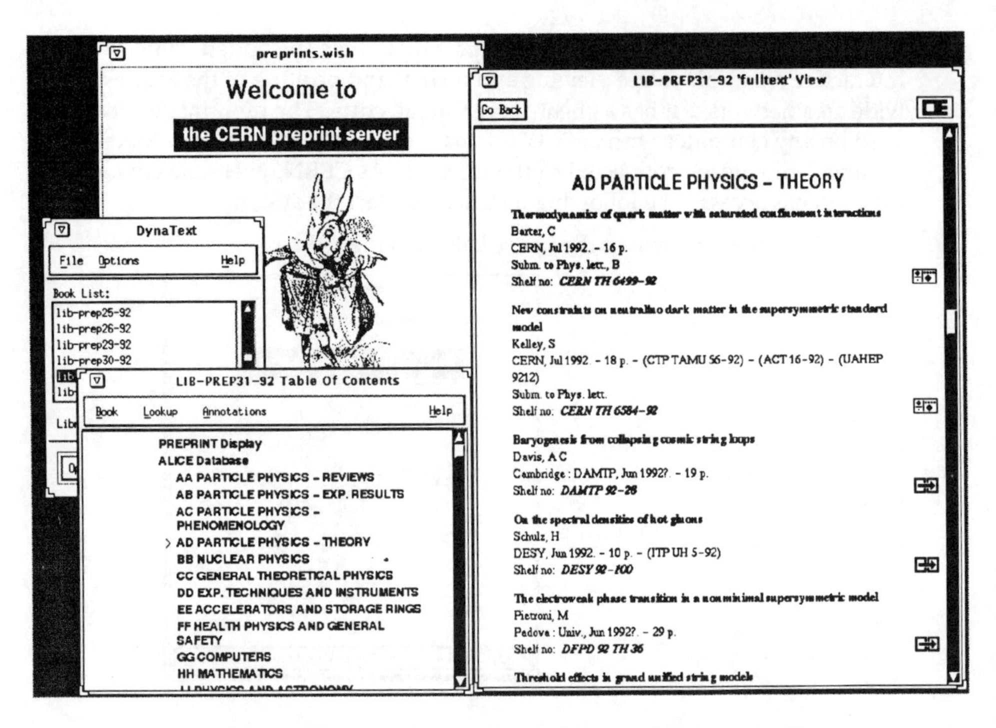

Figure 13.4: The *Dyna*Text version of the weekly preprint list.

This is shown in figure 13.4.

This illustrates the power of SGML: a single source generates two different representations of the same information: one on paper and a second as a hypertext system on a computer screen. 25% of all the documents indexed by the weekly list are available in PostScript, a proportion that will grow rapidly when authors discover the advantages of the system. Preprints not available in PostScript are scanned and stored in compressed format on an optical disk system run by the library. Buttons in the *Dyna*Text version of the weekly list are programmed to start a remote procedure which logs on to this computer and launches a previewer for the desired preprint.

The production of the *Dyna*Text form of the weekly list has been completely automated. Articles, which can be converted to SGML, can also be made available in *Dyna*Text, thus permitting queries about information inside them. The hypertext representation of an article will have buttons for bibliographic references and links to figures, tables and other parts of the document.

The CERN preprint server is not a closed system, but rather based on an architecture which includes the de-facto standards PostScript, X11 and SGML. It allows all forms to co-exist, i.e. paper documents are scanned, TEX, Word and Framemaker documents are converted to PostScript, and SGML documents are made available in hypertext form. *Dyna*Text documents can be grouped together, so searches apply to the whole database. In addition to providing links to files, buttons can be programmed to execute bibliographic searches in Aleph. A system such as this can therefore be used as an alternative user interface to Aleph, and offer full-text and hypertext. Some people dispute the practicality of storing PostScript. However, if all 12,000 preprints of 1992 in PostScript would not have required more than 5 Gbytes of storage space. There are advantages for a large institute to offer these files as a service to its users.

It is not yet clear if hypertext and full-text will extend their utility beyond simple reading aids. See Kircz (1991) for a discussion on how structured documents, SGML and hypertext can improve information retrieval.

Many components of the server are available from the public domain, or come with the Unix operating system. Maintenance is easy and the resources required are reasonable. It demonstrates how a cheap yet powerful documentation system can be built using "open" standards.

13.4.5 Problems integrating bulletin boards with library systems

Not only the number of articles is increasing, but also the number of bulletin boards. None of these systems have been organized by libraries. A coordination will soon be required to manage resources, and to automatically download bibliographic information into databases. Some institutes have stopped sending preprints in paper, but many are not yet ready for a conversion to electronic forms.

Anti-preprinting has not become easier. Sometimes, authors put a preliminary version of their paper on the bulletin board to get feedback, before making it an "official" preprint.

13.4.6 The future role of journals

Traditionally, publishers have four roles: certification (peer review), publishing on paper (in a journal), dissemination (via subscriptions and mail) and archival (in libraries). How should these four areas of activity be transformed when users are setting up networked publication systems such as the ones described above?

Certification

One of the main reasons for the existence of preprints are the delays caused by peer review. Ginsparg (1992) goes a step further: "...the irrelevance of refereed journals to ongoing research in the elementary particle theory community has long been recognized. At least since the mid 1970s, the preprint distribution system has been the primary means of communication of new research ideas and results."

Moreover, there are instances of severe errors in the peer-review system. R. Ernst from the ETH in Zürich, for example, had his paper refused by three reputable journals before being able to publish it (one of them the *Journal of Chemical Physics*). This work obtained him the Nobel prize for chemistry in 1991, and he is now one of the editors for JCP...

Nevertheless, one of the functions of traditional journals is to guarantee a certain level of the quality of the published material. The fact that an article has been screened by an editor specializing in a certain field, and read by a referee considered an expert in that area avoids readers having to waste their time with "polluting noise". Knowing that a manuscript will be read by at least one referee, with necessarily critical eyes, authors mostly avoid sending any insignificant work. It is also more unpleasant to circulate an incorrect preprint than a wrong electronic file which is much easier to destroy. The ideal situation is probably somewhere in between.

Bulletin boards are unselective. The quality of the contributions on the CERN preprint server is guaranteed for institutes such as CERN, SLAC, DESY etc. where papers are internally read and refereed by several people. Contacts between CERN and the libraries of these institutes guarantee that only official publications are used. The quality of the contributions which come from the Los Alamos and other servers is less certain.

A possible improvement, suggested to me by R. Balian (1993) and which can be implemented immediately, would be the requirement that authors include publication details of their article like *"Submitted for publication to (name of journal) on (date)"*. If these details are absent, a note *"Not submitted for publication"* could be added automatically.

Another possibility is to have a bulletin board editor filter the contributions, and some system of electronic refereeing. Preprints and refereed articles would coexist in the same database.

One could go one step further and publish the referee's reports and enable a public discussion of them. Authors that disagree with a referee could leave their article in the database as a preprint. Making the referee's reports public which would provide amusing reading for future Nobel prize winners.

Dating

A problem connected to certification is the need for a clear record of the date of a document because it may be linked to a discovery. Dating electronic documents is prone to fraud since backdating the clock is trivial on computer systems. A solution could be the time-stamping service offered by Bellcore, which provides a computer certificate linking a unique "hash value" from the electronic document to a second value, which is printed every Sunday in the New York Times. This proves that the document could not have been created and submitted to Bellcore for authentication in any other week.

Copyright

To a publisher, the copyright of publishable material is important. Authors, on the other hand, consider their articles as material for the public domain. It is clear that no one will pay money for a printed journal if one can obtain the same articles from a preprint database for free. The copyright mechanism is intended to protect the publisher's livelihood. The free market will probably find a solution to this problem, and it should not be used as an argument to avoid the introduction of new technology.

A role for the publisher could be the following: to take care of the entire database of unrefereed preprints and accepted/rejected articles, without trying to obtain the copyright of the material. A reasonable fee could cover the computer hardware and software resources which are necessary to run the system. Since, however, a major component of journal production, namely the preparation of the printed edition, will disappear, one should see a significant drop in price (a maximum of 2-300K Swiss francs per journal as opposed to the 7500K Swiss francs earned in subscriptions for *Physics Letters B*). Of course such a revolution would also have social and organizational implications inside a publishing company, which I am not in a position to discuss here.

Because of their connections with authors, institutes and building upon the scientific prestige associated with a publication, publishers are in an ideal position to ensure the quality and completeness of a database of scientific material.

Dissemination

The excellent network connections in the academic world ensure faster, cheaper and more widespread distribution of scientific works. If there is a significant drop in prices as suggested above, this will be a particular boon for universities in developing countries who can not afford to buy expensive journals for their libraries.

Archival

There are many advantages in having electronic access to a single database on the network. Provided good retrieval tools exist, it will no longer be necessary for individuals and organizations to keep their own paper copies of articles.

As the database grows, authors will want to make references to articles which exist only in electronic form. The problem of referencing electronic material is yet another topic which is being discussed on the network, and for which a solution should be found. In a hypertext world, one may argue that there is no need for references since they are all replaced by links which immediately call up the cited work.

13.4.7 Internet discovery tools

There exist now a variety of tools which permit the retrieval of information from document servers on the Internet (Schwartz et al., 1992; see also White, chapter 10). W3 (Berners-Lee et al., 1992), for example, has become popular in the physics world. These systems can easily work with bulletin boards via servers. This has been done at SLAC for W3 and the Los Alamos bulletin board.

These tools have varying granularities. For some, the end result of a retrieval is a document or part of it, for others a file, for yet others an electronic message.

13.5 Intelligent Information Retrieval

Supposing all the world's scientific information were available in SGML to a scientist over the network. How would this help the information retrieval process?

To make this process more precise, consider three types of information found in today's libraries, possibly with the help of an automated library system, or an Internet discovery tool:

1. Reference information in dictionaries, encyclopedias, handbooks of mathematical functions etc. This information is easy to retrieve since the kind of questions posed to the database are precise. This is also the kind of information nowadays appearing on CD-ROM. There would be advantages for the publishers of this type of material to use SGML.

2. Information required when learning about a new subject. Usually these are titles of books and articles obtained from colleagues. The subject or skill is learned in the usual way.

3. Information once read or heard, but the reader does not remember where. Citations or data that a reader wants to look up. Which articles to read to keep up to date with progress in the field. This is the kind of information which is published in journals and preprints, and certain books. SGML would have definite benefits to the consumers of this type of information.

Since our database contains the full text of all articles, we can send it a request to look for a given keyword. This would be very costly. SGML, however, can help. Since the titles of the journals, the titles of articles, authors, dates and figure captions are all explicitly marked in the SGML database, we can narrow down the domain of the query. Depending on the granularity of the discovery tool used, the search command can be given to the network or to a server.

13.5.1 Hypermedia

As pointed out earlier, text in SGML is easy to represent in hypertext form. Being able to traverse a document in a non-sequential fashion simulates the way we read paper documents.

Real progress would be made, if, in addition to the various presentation forms the mathematics, figures, histograms and tables, the data themselves would be kept. Readers would be able to choose which form of presentation they prefer, and if required, use the data for whatever purpose.

13.5.2 Limits on information retrieval

SGML is presently not very good for describing tabular material and mathematics. This is because SGML describes tree structures, not two-dimensional objects. A table, for example, can be seen as a set of rows that each have columns, or as a set of columns that each have cells. Both descriptions may be required at the same time, and SGML can not handle this.

A DTD for mathematics is under development (Poppelier et al., 1992), and one for tables will follow later. It is not clear whether one will be able to go beyond queries containing easily identifiable mathematical constructs, since it is impossible to design a language which covers the semantics of the whole of mathematics.

SGML also fails when information on a topic is required of some process, which the author did not have in mind when writing the document. In this case, the title and other text will not contain the information that is used for searching.

13.6 Conclusions

I have reviewed some problems associated with the publication process, the creation and distribution of preprints, and the management of a library.

I have shown that the adoption of SGML could significantly simplify the problems in these three areas. It is unfair to put the burden of this on the authors alone; a firm commitment from the physics community should be made to follow this strategic direction. Librarians and managers of large institutes should play a more active role in this process.

SGML can greatly improve the possibilities for intelligent information retrieval by limiting searches and enabling the use of hypertext technology.

Bulletin boards are also a promising development. A single database solves many problems: there would be no constraint on the author's computer system; there would be no need for anti-preprinting. To automatically extract bibliographic information and to make hypertext, one would still need to convert to SGML.

References

1. Aleph Integrated Library System, Ex Libris, Ltd., Tel Aviv, 1987.

2. ANSI/NISO Z39.59-1988, Standard for electronic manuscript preparation and markup version 2.0, Technical report, EPSIG, Dublin, OH, 1987.

3. Balian, R., "Private communication to the author", *Editor Europhysics Letters*, January 1993.

4. Berners-Lee, T., Cailliau, R., Groff, J.-F. and Pollermann, B., "World Wide Web: the information universe", *Electronic Networking: Research, Applications and Policy*, **2**, 52–58, 1992.

5. Biemesderfer, C., "AASTeX: The foundation of electronic submission", Special insert to the newsletter of the American Astronomical Society, *Electronic Publishing in Astronomy; Projects and Plans of the AAS*, 1992.

6. Charles, D., "Computer fraudsters foiled by the small ads", *New Scientist*, p. 24, 29 February 1992.

7. *Dyna*Text Release 1.5, Electronic Book Technologies, Providence, 1992.

8. Ginsparg, P., "Letter to the editors of Physics Today", *Physics Today*, pp. 13–14 and 100, June 1992.

9. Girvan, R., "The bulletin boards of knowledge", *New Scientist*, 11 July 1992, p. 46.

10. Heck, A., "Results of a desk top publishing survey", *Desk Top Publishing in Astronomy and Space Sciences*, World Scientific, Singapore, pp. 55–63, 1992.

11. ISO/DIS 12083, Information and documentation – Electronic manuscript preparation and markup, Geneva, 1992.

12. Kircz, J., "Rhetorical structure for scientific articles: the case for argumentational analysis in information retrieval", *Journal of Documentation*, **47**, 254–372, 1991.

13. Kircz, J., Research, Publishing and Reading: a Non-Linear Process, Colloquium FOM, 15 June 1992.

14. Oosterhout, J., "TCL: An Embeddable Command Language", *Proceedings USENIX Winter Conference*, January 1990.

15. Poppelier, N.A.F.M., van Herwijnen, E. and Rowley, C.A., "Standard DTDs and scientific publishing", *EPSIG News*, September 1992, 10–19.

16. REVTEX, The American Physical Society, Woodbury, NY, March 1991.

17. Schwartz, M.F., Emtage, A., Kahle, B. and Clifford Neuman, B., "A comparison of internet resource discovery approaches", *Computing Systems 5.4*, Regents of the University of California, 1992.

18. SLAC, Preprints in Particles and Fields, SLAC, PPF-92-39, 1992.

19. The Spires Consortium, Stanford University, Stanford, CA, 1993.

20. Theisen, T., *Ghostview 1.3*, Dept. of Computer Science, University of Wisconsin-Madison, May 1992.

21. Thomas, E., Revised List Processor (LISTSERV), École Centrale de Paris, January 1986.

22. van Herwijnen, E., *Practical SGML*, Kluwer Academic Publishers, Dordrecht, 1990.

23. van Herwijnen, E. and Sens, J.C., "Streamlining publishing procedures", *Europhysics News*, 171–174, November 1989.

24. van Herwijnen, E., Poppelier, N.A.F.M. and Sens, J.C., "Using the electronic manuscript standard for document conversion", *EPSIG News*, **1**, 14, 1992.

25. Weinstein, M., "Everything you wanted to know about PHYZZX but didn't know how to ask", Technical Report SLAC-TN-84-7, SLAC, Stanford, October 1984.

Chapter 14

Astronomical Data Centres from an IIR Perspective

Michel Crézé
Director
Centre de Données de Strasbourg
and
Observatoire de Strasbourg
11, rue de l'Université
F-67000, Strasbourg (France)
Email: creze@simbad.u-strasbg.fr

Why should astronomers care for information retrieval? Should they in fact care at all? Why should they not just wait for general purpose tools developed by IIR professionals to come onto the market? This set of questions inspires another set. What is specific about astronomy? Are Astronomical Data Centres still needed in the era of electronic publishing, CD-ROMs and portable data bases? Several contributors to this volume do emphasize the dramatic growth in volume and complexity of data and information produced by astronomy. Some also stress the fast rise in the publication rate resulting in the same "discoveries" being made three or four times over. Very similar facts can be observed in physics, chemistry or biology. In this regard, there is nothing very specific to any of these disciplines, and the remedy, if any, will probably be a common one.

It is clear, though, that each discipline will have to work with IIR specialists. No tool will be offered for immediate use without an adaptation effort in a particular discipline. Observational data, instrumental know-how, published data, more elaborate knowledge and concepts form an amorphous material. Some effort by knowledgeable scientists is needed to mould this material into appropriate structures, which would make it possible to plug them into general purpose tools such as hypertext. General purpose information structures like SGML are promising, but the implementation of such structures in each specific field will demand major investment. Astronomical Data Centres will probably

A. Heck and F. Murtagh (eds.), Intelligent Information Retrieval: The Case of Astronomy and Related Space Sciences, 193–195.

see their role move from the development of "home brewed" information retrieval tools to the implementation in astronomy of tools developed externally. They must take care of what is specificaly needed by astronomy.

Astronomy has one specificity with respect to e.g. physics: It is always based on an inventory approach. Astrophysicists cannot change the parameters of their experiments, neither can they change the position of the observer with respect to the experiment. The only way around this is to seek another place in the universe where an almost similar experiment occurs with different parameters. Furthermore, the universe as a whole with its time evolution is the basic physical experiment which the astronomer observes. So each piece of the jigsaw puzzle matters. Data cannot be thrown away once they have led to a conclusion. Future observations, and observations at other wavelengths, may at any moment call for re-investigation. Data and information should survive which means that they should remain both accessible and intelligible.

The cleverest database management system is likely to get obsolete on timescales of a few years. The most reliable storage medium may turn to nothing just because a company stops maintaining the necessary read/write device for technological or commercial reasons. (Who can still read punched cards or 7-track tapes?). So long-term information retrieval needs permanent structures able to follow technological evolution and keep the capability of moving the common memory from one medium to the next.

It is not only a matter of volume storage and copying capability. Someone is needed to identify the most suitable data structures. Someone is needed to think of the balance between saving raw data, together with all instrumental information, and saving only carefully chosen subsets of processed data (catalogues and secondary archives). The opposition often made between the two approaches is somewhat artificial: in modern experiments the volume of raw data is so large and the associated documentation so complex that saving additional subsets of calibrated data does not add much to the archiving problem. On the other hand, as the years go by, the effort needed to copy the raw material may result in difficulties. Even if long term archiving is maintained, catalogues and secondary archives will be more readily at hand for most investigations. They would serve at least as pointers to useful raw data.

Most astronomers, at least those who know about complex techniques and observations, are *chromatic*. It would be unwise to use the raw signal of an infrared camera without the help of an expert. On the other hand astronomical objects are basically *panchromatic*. Accretion discs emit X-rays, while surrounding dust and cold gas re-emit infrared radiation. Dust, molecular clouds, ionised and neutral hydrogen, each tracing specific aspects of the interstellar medium, emit and absorb in a wide energy range. Common sense suggests that data would better be stored, managed, structured by people who know about the observing techniques. But this may turn out to be a dramatic limitation to panchromatic investigation. This is not out of any bad will by the specialist, but simply because unless you get some indication that an observation from your current field may be relevant to your investigation, you will never ask for it.

Overspecialization is still more amplified by the proliferation of newsletters through which specialists exchange technical details. Panchromatic communication is likely to be a field where sophisticated IIR tools will not be efficient unless Data Centres working with specialists create the necessary catalogues.

In conclusion, therefore, we see that the evolution of astronomy towards ever greater use of multi-observation, panchromatic data requires IIR as a *sine qua non*. IIR will be of strategic importance to astronomy in the coming years.

Chapter 15

Epilogue

André Heck

Observatoire de Strasbourg
11, rue de l'Université
F-67000 Strasbourg (France)
Email: u01105@frccsc21.bitnet

and

Fionn Murtagh[1]

Space Telescope – European Coordinating Facility
European Southern Observatory, Karl-Schwarzschild-Str. 2
DW-8046 Garching/Munich (Germany)
Email: fmurtagh@eso.org

The following is not a summary of the material treated in this book, but is rather a few salient points. We think they will earn increasing attention in the short term future.

- **Distributed information servers** are increasingly available. They embrace catalogues, yellow page services, electronic journals, and other sources of information. It is to be expected that such servers will soon be availed of in each and every institution.

- It can not be ignored, though, that **adequate resources** – human and technical – are called for, for the establishment and the maintenance of distributed information servers.

[1] Affiliated to the Astrophysics Division, Space Science Department, European Space Agency.

A. Heck and F. Murtagh (eds.), Intelligent Information Retrieval: The Case of Astronomy and Related Space Sciences, 197–198.

- It is not clear at this time if systems such as ESIS and ADS will bear fruit. On the other hand, the integration of information systems into wide-area distributed systems is very much favoured. **Direct access** to information systems also seems to be considerably preferable to clones, unless the latter are explicitly necessitated.
- Databases and archives should be **freely available** for scientific research. They constitute an important social resource.
- Equally important is the availability of an adequate **telecom infrastructure**, – highways and airwaves of the mind.
- **Electronic publishing**, in itself, no longer poses any fundamental technical problem. Delicate points to be solved are essentially **political and social** ones: attitude of, and towards, existing publishers; copyright problems; due protection of electronically published material; personal and collective inertia; adoption of appropriate ethics and an adapted deontology; and so on.
- There is no reason why provision should not be made for systems other than TEX-related ones in electronic publishing. Shifting to **markup languages** will be necessary for all journals and publishers currently recommending the use of TEX-related systems.
- There is a need for wider availability of **automated conversion tools** which link formatting systems and markup languages.
- The role of institutes' **libraries** is fast changing, and with it the work of the librarian. Personal initiative and proper funding are both required, so that the library can become the informational heart of tomorrow's research centre.

Abbreviations and Acronyms

A&A	Astronomy & Astrophysics
A&AA	Astronomy & Astrophysics Abstracts
AAAI	American Association for Artificial Intelligence (USA)
A&AS	Astronomy & Astrophysics Supplement Series
AAS	American Astronomical Society (USA)
ACM	Association for Computational Machinery (USA)
ADASS	Astronomical Data Analysis Software and Systems
ADS	Astrophysical Data System
AI	Artificial Intelligence
AIR	Advanced Information Retrieval
AJ	Astronomical Journal
ALD	Astronomy from Large Databases
AMPTE	Active Magnetospheric Particle Tracer Explorer
ANS	Automatic Navigation System
ANSI	American National Standards Institute (USA)
ApJ	Astrophysical Journal
ApJS	Astrophysical Journal Supplements Series
ARCQUERY	Archive Query
ASCII	American Standard Code for Information Interchange
AXAF	Advanced X-ray Astrophysics Facility
BD	Bonner Durchmusterung
Bitnet	"Because It's Time" Network
BSD	Berkeley Standard Distribution
BSI	Bibliographical Star Index (CDS)
CA	California (USA)
CADC	Canada Astronomical Data Centre
CALS	Computer-aided Acquisition and Logistics Support
Caltech	California Institute of Technology (USA)
CCD	Charge-Coupled Device
CD	Committee Draft
	Compact Disk
CDF	Common Data Format
CDS	Centre de Données de Strasbourg (France)
CERN	Centre Européen de Recherches Nucléaires
CE	Correlation Environment
CfA	Center for Astrophysics (USA)
CFHT	Canada-France-Hawaii Telescope
CGM	Computer Graphics Metafile
CLIM	Common LISP Interface Manager

CLOS	Common LISP Object System
CMS	Conversational Monitoring System
CNRS	Centre National de la Recherche Scientifique (France)
CO	Colorado (USA)
CODATA	Committee on Data for Science and Technology (ICSU)
CPU	Central Processing Unit
CR	Carriage Return
CSI	Catalogue of Stellar Identifications (CDS)
CSO	Computer Services Office (U. Illinois, USA)
DADS	Data Archive and Distribution System
DAL	Data Access Language
DAO	Dominion Astronomical Observatory (Canada)
DBMS	Database Management System
DCF	Document Composition Facility
DEC	Declination
	Digital Equipment Corporation
DESY	Deutsches Elektronen-Synchrotron (Germany)
DIP	Document Image Processing
DIRA	Data Information Retrieval in Astronomy
DIS	Draft International Standard (ISO)
DOS	Disk Operating System
DR	Document Retrieval
DTD	Document Type Definition
DVI	Device Independent
DWIM	Do What I Mean
EARN	European Academic Research Network
ECF	European Coordinating Facility (HST)
Ed.	Edition
	Editor
Eds.	Editors
email	Electronic Mail
EP	Electronic Publishing
EPS	Encapsulated PostScript
EPSIG	Electronic Publishing Special Interest Group (ACM)
ESA	European Space Agency
ESIS	European Space Information System (ESA)
ESO	European Southern Observatory
EXOSAT	European X-ray Observatory Satellite
FITS	Flexible Image Transport System
FOM	Stichting voor Fundamenteel Onderzoek der Materie (Netherlands)
FS	Factor Space

FTP	File Transfer Protocol
GBytes	Gigabytes
GIF	Graphics Interchange Format
GNU	GNU is Not Unix
GSC	Guide Star Catalog (HST)
GSFC	Goddard Space Flight Center (NASA)
GUI	Graphical User Interface
HD	Henri Draper
HEASARC	High-Energy Astrophysics Science Archive Research Center (USA)
H-LAM/T	Human using Language, Artifacts, and Methodology in which he is Trained
HR	Hertzsprung-Russell
HST	Hubble Space Telescope
HT	Hypertext
HTS	Hypertext System
HTML	Hypertext Markup Language
HTTP	Hypertext Transport Protocol
IAU	International Astronomical Union
IBM	International Business Machines
ICPR	International Conference on Pattern Recognition
ICSU	International Council of Scientific Unions
IEC	International Electrotechnical Commission
IEE	Institution of Electrical Engineers (UK)
IEEE	Institute of Electrical and Electronic Engineering (USA)
IIR	Intelligent Information Retrieval
Inc.	Incorporated
ING	Isaac Newton Group (Canary Islands)
INSPEC	Information Services for the Physics and Engineering Communities
I/O	Input/Output
IP	Internet Protocol
IPAC	Infrared Processing and Analysis Center (Caltech)
IR	Information Retrieval
IRAF	Image Reduction and Analysis Facility
IRAS	Infrared Astronomical Satellite
IRIS	Institute for Research on Information Retrieval
IRS	Information Retrieval System
ISBN	International Standard Book Number
ISI	Institute for Scientific Information
ISSN	International Standard Serial Number
ISO	International Standards Organization
IT	Information Technology
IUE	International Ultraviolet Explorer

JCP	Journal of Chemical Physics
km	Kilometer
KMS	Knowledge Management System
LAN	Local Area Network
LDH	Large Dynamic Hyperbase
LF	Line Feed
Lisp	List Processor
Ltd.	Limited
MA	Massachusetts (USA)
Mb	Megabits
MB	Megabytes
MD	Maryland (USA)
MDBST	Multidimensional Binary Search Tree
MI	Medical Imaging
	Michigan (USA)
MIDAS	Munich Image Data Analysis System (ESO)
MMST	Microelectronics Manufacturing Science and Technology
MN	Minnesota (USA)
MNRAS	Monthly Notices of the Royal Astronomical Society
Mpc	Megaparsec
MS	Microsoft
MVS	Multiple Virtual System
NASA	National Aeronautics and Space Administration (USA)
NC	North Carolina (USA)
NED	NASA Extragalactic Database
NGC	New General Catalogue
NISO	National Information Standards Organization (USA)
NJ	New Jersey (USA)
NLS	oN-Line System
NN	Nearest Neighbour
NNTP	Net News Transfer Protocol
No.	Number
NSF	National Science Foundation (USA)
NSFnet	NSF Network
NSSDC	National Space Science Data Center (NASA)
NTT	New Technology Telescope (ESO)
NY	New York (USA)
OCR	Optical Character Recognition

OH	Ohio (USA)
Opt.	Optical
OS	Operating System
OWL	Office Workstation Ltd.
p.	Page
PA	Pennsylvania (USA)
PARC	Palo Alto Research Center
PASP	Publications of the Astronomical Society of the Pacific
PC	Personal Computer
PCES	Physics, Chemistry, and Earth Sciences
PDB	Personal Database
pp.	Pages
PPF	Preprints in Particle and Fields (SLAC)
PPM	Positions and Proper Motions (catalogue)
Proc.	Proceedings
Prolog	Programming in Logic
PS	PostScript
Publ.	Publication
	Publisher
	Publishing
RA	Right Ascension
RAIRO	Revue d'Automatique, Informatique et Recherche Opérationnelle
RAL	Rutherford Appleton Laboratory (UK)
RAM	Random Access Memory
Rep.	Report
RFC	Request For Comments
RFQ	Request For Quotation
RI	Rhode Island (USA)
ROM	Read-Only Memory
ROSAT	Roentgen Satellite
RS	Retrieval System
RTF	Rich-Text Format
SAO	Smithsonian Astrophysical Observatory (USA)
SCAR	Starlink Catalogue Access and Reporting
SDI	Selective Dissemination of Information
SE	Software Engineering
SERC	Science and Engineering Research Council (UK)
SGML	Standard Generalized Markup Language
SIGIR	Special Interest Group on Information Retrieval (ACM)
SIMBAD	Set of Identifications, Measurements, and Bibliography for Astronomical Data (CDS)

SLAC	Stanford Linear Accelerator Center (USA)
SMART	System for Manipulation And Retrieval of Text
SMC	Systems, Man, and Cybernetics
SMM	Solar Maximum Mission
SPIKE	Science Planning using Intelligent Knowledge-based Environment
SQL	Structured Query Language
SSDB	Scientific and Statistical Database
STARCAT	Space Telescope Archive and Catalogue
ST-ECF	Space telescope - European Coordinating Facility (HST)
STELAR	Study of Electronic Literature for Astronomical Research (NASA)
STI	Scientific and Technical Information
STScI	Space Telescope Science Institute
STSDAS	Space Telescope Science Data Analysis System
Suppl.	Supplement
TCL	Tool Command Language
TCP	Transmission Control Protocol
TFS	Table File System
TIES	The Interactive Encyclopaedia System
TIFF	Tag Image File Format
TM	Trade Mark
TMC	Thinking Machines Corp.
TN	Technical Note
Trans.	Transactions
TTY	Teletype
TV	Television
UDI	Universal Document Identifier
UIF	User Interface
UIT	Ultraviolet Imaging Telescope
ULDA	Uniform Low Dispersion Archive (IUE)
UKS	UK Satellite
URL	Universal Resource Locator
VAX	Virtual Address Extension
VERONICA	Very Easy Rodent-Oriented Net-wide Index to Computerized Archives
VLBI	Very Large Baseline Interferometry
VM	Virtual Machine
VMS	Virtual Memory System
Vol.	Volume
WAIS	Wide Area Information Server
WAN	Wide-Area Network
WDC	World Data Centre

WG	Working Group
WGAS	Working Group on Astronomical Software (AAS)
WI	Wisconsin (USA)
WIMP	Window - Icon - Menu - Pointer
WSRT	Westerbork Synthesis Radio Telescope (Netherlands)
WWW	World-Wide Web
WYSIWYG	What You See Is What You Get
W3	World-Wide Web

Index

A&A, 60, 139, 199
A&AA, 6, 27, 54, 199
A&AS, 60, 199
AAS, 138, 199
Abell catalogue, 23
Acta Astronomica, 7
Adorf, H.-M., 49–80, 54, 69
ADS, 6, 22, 27, 54, 69, 136, 148, 149, 198, 199
ADS Abstract Service, 21, 22, 26
AJ, 60, 199
Akscyn, R.M., 96, 99
Albrecht, M.A., 50, 52, 54, 70–74, 76, 135–152, 154, 170
Albrecht, R., 70, 75
Aleph, 179, 183, 189
Alfvén speed, 167
Alleyne, H.St.C., 171
AMPTE-UKS, 159, 199
André, J., 99–102
Andreas, G., 100
Anklesaria, F., 119–125
anonymous ftp, 104, 122, 131, 181
Ansari, S.G., 170, 171
anti-preprint, 176, 178, 179, 180, 185
ApJ, 60, 139, 200
ApJS, 60, 200
archie, 103–111, 119, 122, 132
ARCQUERY, 53, 141, 199
Arnd, T., 72
artificial intelligence, 4
Association of American Publishers, 177
Astronet, 51
Astronomical Journal, 139
Astronomischer Jahresbericht, 6
Astronomy & Astrophysics Abstracts, *see* A&AA
Astronomy & Astrophysics Supplement Series, 7
Astronomy and Astrophysics, *see* A&A
Astronomy and Astroyphsics Monthly Index, 58
Astrophysical Data System Abstract Service, *see* ADS Abstract Service
Astrophysical Journal, *see* ApJ
Augment, 88
authoring system, 89
AutoCAD, 67
automatic indexing, 13

Baeza-Yates, R., 30, 46, 83, 99, 101
Balian, R., 186, 189
bar codes, 7
Barnes, J., 28, 151, 171
Barylak, M., 151
Baskett, F., 46
BD catalogue, 145
Belew, R.K., 13, 19
Belkin, N.J., 9–20, 50, 71
Bellcore, 178, 186
Benacchio, L., 51, 71
Bennett, J., 73, 151
Bentley, J.L., 32, 34, 43, 45, 46
Benvenuti, P., 51, 71, 76
Berners-Lee, T., 69, 71, 111, 130, 133, 187, 189
Bernstein, M., 90, 99
best match, 11, 29, 95
Betz, D., 64
Bezdek, J.C., 48
Bibliographical Star Index, *see* BSI
BIBTeX, 180
Bicay, M.D., 151
Biemesderfer, C., 151, 171, 178, 189
Biennier, F., 98, 99
Billings, S.A., 170
BinHex, 122
Biswas, G., 48, 98, 99
Blelloch, G.E., 68, 76
Bloech, J., 83, 99
Bocay, M.D., 73
Bodini, A., 71
Bogaschewsky, R., 69, 71, 81–102
boolean search, 11, 55–57, 60, 94, 95

Bothun and Mould, 23
Brackeen, D., 66, 72
Brajnik, G., 13, 18, 19
Brissenden, R.J.V., 28, 151
Broder, A.J., 34, 45
Brotzman, L., 76
BRS, 58
Brugel, E.W., 74
Bruza, P.D., 94, 100
BSI, 6, 199
Buckley, C., 12, 20
bulletin boards, 119, 181
Bunyip Information Systems, Inc., 103, 106
Burkhard, W.A., 38, 46
Busch, E.K., 54, 71
Bush, V., 88, 100

Cailliau, R., 111, 133, 189
Canada-France-Hawaii Telescope, *see* CFHT
Carando, P., 89, 100
Carmel, E., 94, 100
Carpenter, B., 50, 72
Catalogue of Stellar Identifications, *see* CSI
Cawsey, A., 13, 19
CCD, 22, 199
CD-ROM, 188, 193
CDF, 166, 199
CDS, 145, 149, 199
CFHT, 137, 199
CGM, 167, 199
Chambers, J.M., 64, 72
Chang, S.-K., 68, 72
Charles, D., 178, 190
Chebyshev metric, 37
Chen, S., 170
Cheng, Y., 50, 72
Chevy, D., 58, 72
Chignell, M.H., 68, 72, 74, 82, 92–94, 96, 102
Christian, C.A., 51, 72, 151
Ciarlo, A., 151, 154, 170, 171
class discovery, 106, 107
client-server model, 58, 129
Clifford Neuman, B., 111, 190
CLIM, 77, 199
Clitherow, P., 90, 100
CLOS, 77, 200
cluster analysis, 41
Codd, E.F., 161, 171
cognitive
  overhead, 86
  overload, 50
  task analysis, 14
Cohen, Y., 100
cold dark matter, 27
Collins Dictionary, 92, 100
Common Data Format, 166
compuscript, 6, 177
computer graphics metafile, 167
computer pen writing, 7
CONF, 58
Conference Papers Index, 58
Conklin, J., 86, 87, 89, 100
Consens, M.P., 95, 100
Cool, C., 19
copyright, 186
Corbin, B.G., 58, 72
Cornish, M., 72, 77
correlation environment, 138, 153–171
Cousins, S.B., 96, 97, 101
Cove, J.F., 93, 100
Cox, R.T., 50, 72
Crabtree, D.R., 137, 151
Crawford, D., 72
Crézé, M., 193–195
Croft, W.B., 11, 13, 18–20, 39, 40, 46, 50, 71, 97, 100
Crouch, C.J., 100
Crouch, D.B., 98, 100
CSI, 6, 200
CSO name server, 123, 124
Current Contents, 54–56, 58, 66, 68
cyberspace, 7

D2 Software Inc., 63
DADS, 44, 68, 200
DAL, 62, 200
Daniels, P.J., 13, 19
DAO, 137, 200
Dasarathy, B.V., 30, 46
Data Access Language, 62
Data Archive and Distribution System, *see* DADS
data centres, 193–195
Date, C.J., 52, 72
Davenhall, A.C., 51–53, 72
Davenport Group, 86
DBMS, 50, 57, 60, 200
Debye length, 167
Deerwester, S., 29, 46
Delannoy, C., 32, 46

Dempster-Shafer theory, 98
Denning, P., 50, 72
DESY, 180, 200
Deutsch, P., 103
Di Gesù, V., 73, 75
Dialog, 56, 58
Dimitrof, D., 72
DIRA, 51
Dobson, C., 72
document
    learning, 11
    retrieval system, 83
Document Type Definition, 177
Dominion Astronomical Observatory, *see* DAO
Donzelli, P., 170, 171
DTD, 177, 189, 200
Dubois, P., 72, 151
Dumais, S.T., 11, 20, 46
Durand, D., 74, 151, 152
DWIM, 66, 200
Dynatext, 180, 183, 184, 190

Eastman, C.M., 36, 46
Egan, D.E., 17, 19
Egret, D., 50, 51, 53, 71–74, 76, 135–152
Eichhorn, G., 28, 74
electronic junk, 50
electronic publishing, 1, 6, 7, 138, 200
Ellis, D., 82, 90, 98, 100
Emacs, 67, 69
Emtage, A., 103–111, 190
EndLink, 57
EndNote, 56–58, 68
Engelbart, D.C., 88, 100
Envision, 99
ESA/IRS, 58
ESIS, 6, 14, 22, 54, 69, 136, 138, 146–149, 154, 170, 198, 199, 200
ESO archive, 52, 58, 146
ESO-MIDAS, 51, 147
Euclidean distance, 31
European Physical Society, 177, 178
exact match, 11, 29, 95
Excel, 61, 62, 66, 67, 69, 78
EXOSAT, 51, 52, 137, 143, 144, 147, 200
experimental databasing, 155
expert system, 4

factor space, 21, 26, 27, 29
Fekete, G., 43, 46
Ferris, A., 74
file transfer protocol, *see* ftp
Finger, 119
Finkel, R.A., 46
FITS, 6, 142, 143, 146, 200
Foltz, P.W., 11, 20
Fox, E., 20
Fraase, M., 88, 89, 96, 100
Frakes, W.B., 30, 46, 99
Friedman, J.H., 33, 34, 37, 43, 45, 46
Frisse, M.E., 96, 97, 101
ftp, 104, 108, 119, 122, 124, 131, 181, 201
Fuhr, N., 95, 96, 101
Fukunaga, K., 38, 46
fulltext retrieval system, 83, 89
Fullton, J., 113–117
Furnas, G.W., 11, 20, 46
Furuta, R., 90, 102
fuzzy
    logic, 170
    search, 45

Galliers, J., 19
Gass, J., 76
Gaudet, S., 74, 152
genetic algorithms, 170
Gershenfeld, N.A., 48
Ghostview, 183
Giaretta, D.L., 154, 170, 171
Ginsparg, P., 181, 185, 190
Giommi, P., 14, 20, 52, 73, 77, 137, 138, 143, 144, 148, 151, 152, 154, 170, 171
Giovane, F., 76
Girvan, R., 181, 190
Glaspey, J, 151
Gloor, P.A., 89, 101
Gomez, L.M., 20
Gonnet, G.H., 83, 99, 101
Good, J.C., 54, 74, 77, 148, 152
Gopher, 58, 111, 119–125, 128, 131, 132
Goppert, R., 39, 47
gravitational lens, 27
grey literature, 5
Griffiths, A., 42, 46
Grits, D., 72
Groff, J.-F., 111, 133, 189
Grosbøl, P., 52, 71, 74, 142, 150–152
Grosky, W.I., 30, 46
Groucher, G., 169, 171
GSC, 149, 201

GUI, 65, 67, 98, 141, 149, 201
Guida, G., 19
Guide, 90
Guide Star Catalog, 142
Guivarch, M., 99

H-LAM/T, 88, 201
Hájek, P., 50, 73
Hafner, A., 67, 73
Halasz, F.G., 92, 101
Hamming metric, 37
Hammwöhner, R., 93, 101
Hanisch, R.J., 28, 50, 51, 73, 136, 151
Hapgood, M.A., 71, 153–171
Harman, D., 11–13, 20
Harris, A.W., 52, 77
Harshman, R., 46
Hauck, B., 1–2
Hayes, P., 86, 101
HEASARC, 137, 201
Heck, A., 2, 3–8, 20, 28, 42, 47, 70–75, 136, 138, 148, 150–152, 177, 180, 190, 197–198
Helou, G., 49, 52, 53, 58, 73, 137, 141, 145, 151
hierarchical clustering, 41
Hipparchus, 1
Hodgson, M.E., 38, 46
Hofmann, M., 88, 101
Hoppe, U., 81–102
Horn, B.K.P., 77
HR catalogue, 145
HST, 1, 6, 22, 44, 51, 52, 68, 80, 136, 201
HTML, 132, 201
HTTP, 131
Hubble constant, 25
Hubble Space Telescope, *see* HST
humbers, 89
Hurst, A.J., 61, 75
HyperCard, 62, 69, 96
hypermedia, 4, 69, 188
hypertext, 4, 69, 81–102, 127, 128, 180, 183
Hypertext Markup Language, *see* HTML
hypertext transport protocol, *see* HTTP
HyperTIES, 96
HyTime, 86

IAU, 1, 201
IDL, 51, 52, 64, 75, 159, 169, 171
image restoration, 49
indexing, automatic, 13
infostructure, 106

ING, 141, 201
Ingres, 62
Ingwersen, P., 11, 12, 18, 20
INSPEC, 54, 58, 68, 201
instance
    access, 106
    location, 106, 107
Institute of Scientific Information, *see* ISI
Intermedia, 96
International Astronomical Union, *see* IAU
International Ultraviolet Explorer, *see* IUE
Internet Gopher, *see* Gopher
IRAF, 117, 201
IRAS, 143–145, 147, 201
IRS, 58, 81, 82, 201
Isaac Newton Group, 137
    archives, 51, 53
ISI, 55, 201
ISO, 177, 201
IUE, 115, 136, 138, 141, 142, 145, 201

Jaccard similarity coefficient, 31
Jarke, M., 53, 73
Jasniewicz, G., 151
Jenkner, H., 142, 152
Johnson, D., 73, 77
Johnston, M.D., 22, 28, 68, 70, 73
Journal of Chemical Physics, 185
Justus, 90

k-d tree, 33
Kahle, B., 107, 111, 190
Kahn, P.D., 127, 133
Kamgar-Parsi, B., 39, 46
Kanal, L.N., 39, 46
Kantor, P., 11, 20
Karakashian, T., 28
Kashyap, R.L., 50, 72
Keller, R.M., 38, 46
Khoshafian, S., 74
Kimberley, R., 42, 47
Kipp, N.A., 101
Kircz, J., 176, 185, 190
Kittler, J., 37, 47
KMS, 96, 202
Kohonen, T., 33, 47
Kovalsky, D., 76
Kraft, D.H., 30, 47
Krol, E., 50, 58, 73
Kuhlen, R., 89, 92, 93, 101

Kurtz, M.J., 21–28, 74

La Palma archives, 51, 53
Lamport, L., 147, 152
Landauer, T.K., 20, 46
Landsman, W.B., 51, 52, 64, 73
Large Dynamic Hyperbases, *see* LDH
latent semantic indexing, 29
LDH, 94, 202
Lepine, D.R., 171
lex, 53
Liddy, E.D., 18, 20
Lindler, D., 64
linked views, 66
Lisp-Stat, 64–66, 68
Listserv, 181
Locke, 4
Loral 1024 chip, 23
Lotus 1-2-3, 61
Lubinsky, D.J., 64, 73
Lucarella, D., 97, 101
Lucy, L.B., 50, 73

Mackraz, B., 72
MacSpin, 63, 66, 68
Madore, B.F., 73, 151
Malina, R., 72
Manber, U., 83
Marchetti, P.G., 19
Marcus, A., 65, 69, 74
Marimont, R.B., 37, 47
markup language, 146, 198
Matlab, 169
McAleese, R., 92, 93, 101
McCahill, M., 119–125
McCracken, D.L., 99
McGill, M.J., 42, 48, 82, 95, 98, 101
McGlynn, T., 75
McHenry, W.K., 100
MDBST, 33–35, 203
Mehrotra, R., 30, 46
memex, 88
Mendelzon, A.O., 95, 100
metadata, 142, 162
Michalewicz, Z., 50, 74
Micó, L., 38, 47
Microsoft Excel, 61, 62, 66, 67, 69, 78
Midas, 130, 132
Miyamoto, S., 45, 47
MNRAS, 60, 202
Moon, D., 76
Morariu, J., 96, 102
Mount Pinatubo, 22
Muller, M., 100
multidimensional binary search tree, 33
multidimensional scaling, 33
multimedia, 85
Murray, S.S., 28, 54, 74
Murtagh, F., 2, 3–8, 20, 28, 29–48, 50, 51, 70–75, 136, 150–152, 197–198
Myaeng, S.H., 18, 20

Nanni, M., 51, 71
Naranjo, M., 48
Narendra, P.M., 38, 46
Narmax, 170
NASA Scientific and Technical Information Branch, 26
NASA-STI, 26
nearest neighbor, 30
NED, 52, 53, 58, 137, 141, 145, 202
Nelson, T.H., 88, 101, 127, 133
Network News Transfer Protocol, *see* NNTP
neural networks, 13, 170
New Technology Telescope, *see* NTT
Newcomb, S.R., 86, 101
Newcomb, V.T., 101
NGC 4151, 138
Nielsen, J., 69, 74, 89, 101
Niemann, H., 39, 47
NLS, 88, 202
NNTP, 119, 131, 202
Norvig, P., 74, 77
Nousek, J.A., 74
NSFnet, 104, 119, 202
NTIS, 58
NTT, 136, 137, 202
Nyquist theorem, 161

object orientated system, 78
Ochsenbein, F., 52, 74, 75, 152
Oddy, R.N., 13, 20, 48
Oncina, J., 47
Oosterhout, J., 183, 190
Oracle, 62
ORBIT, 58
Ossorio, P.G., 26, 28
OWL, 90, 203
Ozsoyoglu, G., 53, 74
Ozsoyoglu, Z.M., 53, 74

Page, C.G., 51, 74
Paliwal, K.K., 38, 47
panchromatic astronomy, 194
Parsaye, K., 68, 72, 74
partial match, 29, 57, 95
Pascual, J., 71
Pasian, F., 52, 74, 76, 136, 141, 151, 152
PASP, 60, 203
PCES, 55, 203
Penny, A.J., 171
Pepper, J., 86, 101
Péron, M., 51, 74, 148, 152
Perry, S.A., 41, 47
personal database, 148
PHYS, 58
PHYZZX, 178
Piazza, J., 76
PICT, 60
Pinon, J.-M., 99
Pirenne, B., 52, 70, 74, 136, 137, 141, 150, 152
Pollermann, B., 111, 133, 189
Pomphrey, R., 54, 74
Ponz, J.D., 51, 75
Poppelier, N.A.F.M., 178, 189, 190
Postel, J., 104, 111
postscript, 60
PPM, 149, 203
precision (and recall), 12, 95
Preparata, F.P., 43, 47
preprint server, 182
Press, L., 50, 58, 75
Pro-Cite, 56
probabilistic retrieval model, 96
Prospero, 111
publish or perish, 5, 173, 174

quadtree, 43, 44
query by example, 53
query environment (ESIS), 14

Raimond, E., 53, 75, 137, 141, 152
Ramasubramanian, V., 38, 47
Rampazzo, R., 70
Read, B.J., 171
recall (and precision), 12, 95
RECON, 54, 58, 60, 149
redshift survey, 23
Reece, S., 19
reference retrieval system, 83
Regener, P., 170
relevance feedback, 11, 12, 24, 60, 68, 96, 98
resource discovery, 50
restoration, image, 49
REVTEX, 178, 190
Rey-Watson, J.M., 75, 142, 152 *See also* Watson, J.M.
Reynolds, J., 104, 111
Richardson, B.H., 50, 75
Richetin, M., 37, 48
Richmond, A., 52, 71, 74, 75, 141, 152, 170
Riecken, D., 100
Rives, G., 48
Rizk, A., 99–102
rmail, 60
Robinson, L.A., 46
Rohlf, F.J., 33, 48
Romelfanger, F., 75
ROSAT, 52, 204
Rosenthal, D.A., 67, 75
Roth, E.M., 14, 20
Rothrock, M.E., 75, 77
Rowley, C.A., 190
rule-based system, 78
Rumble, J.R., 50, 75
Russo, G., 52, 71, 75, 170

Sajeev, A.S.M., 61, 75
Salton, G., 11, 12, 20, 30, 42, 48, 82, 95, 98, 101
Samet, H., 43, 48
Sammon, J.W., 33, 48
SAO, 145, 203
Saracevic, T., 11, 20
SCAR, 52, 53, 203
Schlechter, G., 101
Schmitz, M., 73, 151
Schreiber, T., 33, 48
Schreier, E., 51, 76
Schwartz, M.F., 106, 111, 187, 190
Scientific and Technical Information, *see* STI
SciSearch, 58
SDI, 4, 203
search algorithms, 29–48
Seeger, T., 19
selective dissemination of information, *see* SDI
Semmel, R.D., 67, 76
Sens, J.C., 177, 180, 190
Sesame, 145, 149
SGML, 86, 139, 143, 147, 177, 180, 183, 184, 188, 189, 193, 203

Shalit, A.M., 76
Shamos, M.I., 43, 47
Shapiro, M., 37, 38, 47, 48
Shepp, L.A., 50, 76
Shneiderman, B., 96, 102
Shoshani, A., 50, 76
Shustek, L.J., 46
SIMBAD, 1, 6, 51, 53, 58, 137, 141, 142, 145, 203
Sipelstein, J.M., 68, 76
SMART, 98, 204
Smeaton, A.F., 41, 48
Smith, F.J., 50, 75
SMM, 137, 204
Solar Maximum Mission, *see* SMM
Sparck Jones, K., 19, 96, 102
speech recognition, 7
sphere quadtree, 43, 44
SPIKE, 22, 204
Spires, 179, 190
spreadsheet, 60, 78
SQL, 52, 53, 62, 67–69, 155, 163, 204
SSDBMS, 50
ST-ECF archive, 58
Stallman, R., 67, 69, 76
STARCAT, 53, 67, 70, 79, 80, 137, 141, 149, 204
STARMAIL, 141
StarServer, 149
StarView, 68
statistical factor space, 21, 26, 27, 29
Steele, G.L., 76, 77
Stein, A., 19
Steinacker, I., 83, 102
STELAR, 58, 115–117, 204
Stern, C.P., 28
STI, 26, 204
STN, 58
Stokke, H., 151, 170, 171
Stoner, J.L., 28, 74
Stotts, P.D., 90, 102
Strömgren
    photometry, 22
    system, 22
Streitz, N.A., 99–102
Structured General Markup Language, *see* SGML
STSDAS, 51, 204
Su, L., 12, 20
Subramanian, V., 45, 48
Sybase, 62

Tagliaferri, G., 52, 73
Tarasko, M.Z., 50, 76
Tasso, C., 19
TCL, 183, 204
telnet, 104, 108, 119, 123, 124
Tesler, L., 76, 77
TeXpert, 180
Theisen, T., 183, 190
Thiel, U., 19
Thinking Machines Corp., 58
Thomas, E., 181, 190
Thompson, R.H., 13, 19
Tierney, L., 64, 76
TIFF, 60, 204
time allocation committee, 23
tkWWW, 132
TMC, 58, 204
Tognazzini, B., 72
TOPBASE, 138
topological mapping, 33
Torrente, P., 170, 171
Touretzky, D.S., 76, 77
traveling salesman problem, 30
triangular inequality, 37
trixels, 44
Tully-Fisher, 23
Turtle, H., 11, 13, 18, 20, 97, 100

UDI, 131, 204
UIF, 50, 52, 54
UIT, 52, 64, 204
ULDA, 142, 145
uncertainty reasoning, 50
Universal Document Identifier, *see* UDI
Universal Resource Locators, *see* URL
URL, 129–132, 204
Usenet, 105, 109
uuencode, 122

van Dam, A., 65, 69, 74
van Herwijnen, E., 147, 152, 173–191
van Rijsbergen, C.J., 11, 20, 41, 48, 95, 98, 102
Van Steenberg, M.E., 58, 76
Vardi, Y., 50, 76
Vassiliou, Y., 53, 73
vector space model, 95
Verne, J., 7
Veronica, 122, 204

Vickery, A., 16, 19
Vidal Ruiz, E., 38, 47, 48
video-optical devices, 5
Viola, 132

W3, *see* WWW
WAIS, 53, 58–60, 67, 68, 70, 79, 80, 107, 108, 113–117, 119, 122, 128, 131, 132, 149, 204
Walker, S.N., 170, 171
Walsh, B.C., 93, 100
Wamsteker, W., 136, 143, 151, 152
Wang, P.S., 67, 76
Warnock, A., 60, 76
Waterworth, J.A., 82, 92–94, 96, 102
Watson, J.M., 28, 54, 58, 76 *See also* Rey-Watson, J.M.
Watson, R.M., 58
Webber, R.E., 43, 48
Weerawarana, S., 67, 76
Weide, B.W., 45
Weigend, A.S., 48
Weinstein, M., 178, 191
Weiss, J.R., 54, 77, 148, 152
Weiss, S.F., 36, 46, 48
Wenger, M., 72, 151
Wersig, G., 19
Westerbork archives, 51, 53
Westerbork Synthesis Radio Telescope, 137
white pages, 104
White, B., 127–133, 187
White, N.E., 52, 77, 137, 143, 152
whois, 119
Willett, P., 41, 46, 47
Williams, D., 50, 72
Wilson, E., 90, 102
Winston, P.H., 77
Wong, H.K.T., 50, 74, 76
Woods, D.D., 14, 20
Woolliscroft, L.J.C., 170, 171
Worrall, D.M., 151, 171
WRST, 141
Wu, H., 20
Wu, S., 83
Wu, X., 73, 151
WWW, 58, 99, 111, 127–133, 187, 205

X500, 124
Xanadu, 88
Xerox PARC, 65
XLISP, 64
XMosaic, 132

Yan, C.W., 72
Yankclovich, N., 96, 102
Yao, A.C., 45
yellow pages, 104
Yoder, E.A., 99
Yovits, M.C., 46
Yunck, T.P., 37, 48

Zimmermann, H.U., 52, 77
Zwicky catalogue, 23